Wojciech Kopczyński is extraordinary professor at the Institute of Theoretical Physics at Warsaw University. He was born in Toruń in 1946 and graduated from the Physics Department of Warsaw University in 1969. In 1973 he obtained his Ph.D. degree in physics under the supervision of Andrzej Trautman. He was made assistant professor at the Physics Department of Warsaw University in 1987. He is author of more than thirty publications on the theory of gravitation and mathematical methods in physics. In 1973–74, funded by the French Government, he researched at Collège de France. In 1979–81 and again in 1986, he was a Humbold Foundation scholarship holder at the University of Cologne. He is married with two sons.

Spacetime and Gravitation

Spacetime and Gravitation

Wojciech Kopczyński and **Andrzej Trautman**
Institute of Theoretical Physics,
Warsaw University

A Wiley—Interscience Publication

JOHN WILEY & SONS
Chichester · New York · Brisbane · Toronto · Singapore
PWN—POLISH SCIENTIFIC PUBLISHERS
Warszawa

Revised translation of the Polish original *Czasoprzestrzeń i grawitacja*, published in 1984 by Państwowe Wydawnictwo Naukowe, Warszawa

Chapters 1, 7, 9, 11–19 translated by *Jerzy Bałdyga*, Chapters 2–6, 8, 10 by *Antoni Pol*

Graphic design by *Zygmunt Ziemka*

Co-published by John Wiley & Sons Ltd, Baffins Lane, Chichester, West Sussex, PO19 1UD, UK.

Library of Congress Cataloging-in-Publication Data:
Kopczyński, Wojciech.
[Czasoprzestrzeń i grawitacja. English]
Spacetime and gravitation / Wojciech Kopczyński and Andrzej Trautman ; [translated by Jerzy Bałdyga, Antoni Pol].
p. cm.
"Revised translation of the Polish original Czasoprzestrzeń i grawitacja" — CIP t.p. verso.
Includes bibliographical references and indexes.
ISBN 0-471-92186-6
1. Relativity (Physics). 2. Geometrodynamics. 3. Gravitation.
I. Trautman, A. (Andrzej) II. Title.
QC 173.55.K6713 1991 90-27813
530.1'1 — dc20 CIP

British Library Cataloguing-in-Publication Data:
Kopczyński, Wojciech
Spacetime and gravitation.
I. Title II. Trautman, Andrzej
530.11

ISBN 0-471-92186-6

Printed in Poland by D. N. T.

Contents

Foreword

In 1969, at the initiative of Witold Nowacki, then vice-President of the Polish Academy of Sciences, there was organized a series of lectures on "The Achievements of Polish Science". After some time, to continue and extend these conferences, an Open University of the Polish Academy of Sciences was established. Within the framework of the first series of these conferences, one of us (A. T.) gave a course of lectures entitled "The Relativity Theory". The text of these lectures was then extended by the other author (W. K.) and published in 1971 by the Ossolineum. The interest with which this small book met, encouraged us to prepare on its basis the present text. We kept in it—partly, at least—the original, colloquial language of the lectures.

In this little book, we stress above all the conceptual side of the special and general relativity theories. We pay much attention to the geometrical aspects of this theory. This book can serve as a text for those who want to learn the fundamentals of the special and general theories of relativity. It does not, however, offer an exhaustive treatment of these theories. Therefore, a reader who wishes to obtain a more profound knowledge of the theory of relativity should also consult the literature listed at the end of the book.

We believe that this book should be accessible to readers who have had a one-year course in mathematics and physics at the science depertments of universities or at technical universities. We dedicate this book above all to students.

CHAPTER 1

Introduction

The contemporary science of space, time and gravitation, commonly called the relativity theory, emerged in 1905–1916. Although more than half a century has passed since then, it is still met with considerable interest and even fascination. What are the reasons for this. The first, probably the most important, reason is that the relativity theory is concerned with the fundamental notions which we all come across and about which we all have certain ideas, namely the notions of time and space. We should also consider, in this respect, the personality of the founder of this theory: Einstein was an extraordinary person, an eccentric with unusual behaviour; his life and character excite human imagination and encourage us to take interest in his theory.

The very period of time, when relativity theory was created, also played some role: those were the years when new technologies were being developed, electric energy was coming into widespread use, the first aeroplanes appeared and the radio epoch began. The technological achievements contributed to convincing people about the significance of science even in everyday life. At the same time, there were then not so many new inventions that it would be impossible to stop and think about them. Now, scientific discoveries and new technological advances are so frequent that something like the landing of men on the Moon is required to excite wider interest in, and admiration for, the work of engineers and scientists.

Certain factors, which we can call subjective, have also had their effect. We mean by that the activity of popular science writers, quite often physicists themselves, who set about with particular zeal to divulge, and at times to vulgarize, the theory of relativity. Why did that happen? Probably because by speaking of relativity theory we can "astound the bourgeois". We have in mind here the possibility, which was quite frequently made use of, to present the conclusions of relativity theory so as to impress on the listeners the idea that all their previous notions of space and time were mistaken, and that the true picture of the world is accessible only to a small group of the initiated. This approach to the popularization of relativity theory, on the one hand, resulted in creating a vivid interest and numerous discussions of the subject; on the other, it led to the situation that even today we meet people who reject the whole of Einstein's theory and try to turn physics back to the pre-relativistic period.

In the initial period of its history, the theory of relativity was particularly susceptible to popularization, since the experimental material which confirmed it, was rather scarce and could be described in a way accessible to a layman. It was then sufficient to discuss the Michelson–Morley and Trouton–Noble experiments. Clearly, things have now changed and there are many direct and indirect precise tests of the special relativity theory; but there is no universal awareness of the new situation. Many people still believe that relativity theory is beyond a layman's comprehension, its results are paradoxical, while only measurements of the speed of light provide the experimental basis.

We shall not be concerned here with such formulations as: according to Einstein everything is relative, time is imaginary and matter can transform into energy. These apparent wisdoms have emerged on the margin of science and have nothing in common with what is asserted in relativity theory understood as part of physics.

We shall discuss instead the geometry of the different models of time and space, paying attention to similarities and differences between Newtonian and Einsteinian models [51, 52, 53]. We shall pay much attention to problems which find their way into popular discourses and are quite frequently represented incorrectly. What we mean is, e.g. the "twin paradox", the problem of visibility of the relativistic length contraction [23] and the question of the existence of motions faster than light.

Despite what has at times been said on the subject, we should not consider the special theory of relativity as being the creation of Einstein alone, independent of the development of physics at the turn of the 19th and 20th centuries. The emergence of this theory was made possible by the developments in electrodynamics, while its germs could be found in the work of Lorentz, Fitzgerald, Larmor and Poincaré. Of the earlier discoveries, without which we could not think of relativity theory at all, we should mention the first attempt at evaluating the speed of light made by the Danish astronomer Olaf Römer in about 1675. Observing the motion of one of Jupiter's moons, Römer calculated the period of its motion and then found that at a moment when the Earth was farther away from Jupiter, the eclipse occurred more than 10 minutes later than he had predicted from observations of eclipses when the Earth was closer to Jupiter. Hence, he drew the conclusion that light propagated at a finite speed. Since he did not know the exact dimensions of the Earth's orbit, he underestimated this speed. If one were to use the parameters of the Earth's orbit determined later, Römer's calculations would give a speed of light of 310 000 km/sec, an outstandingly precise result.

Of importance was the discovery in 1728 by Bradley of the phenomenon of aberration. The phenomenon consists of the apparent displacement of fixed stars during the Earth's motion around the Sun. In contrast to the parallactic displacements of stars, caused by changes in the direction from which we observe stars, brought about by changes in the Earth's position, aberration is related to changes in the Earth's velocity; it resembles the phenomenon of the formation of oblique rain drops on the window panes of a travelling train. In 1804, T. Young explained the phenomenon of aberration on the ground of the wave theory of light, assuming that light consists of the propagation of vibrations in a stationary ether. If, however, the Earth and bodies lying on it move with respect to the ether, then the light refraction coefficient, and, accordingly, the focal length of a lens, should depend on whether the lens is moving towards the light source (star) or away from it. D. Arago's (1818) observations did not exhibit this phenomenon, and, although, they were probably not accurate enough to settle the case, they encouraged A. Fresnel to look for a theoretical explanation. Fresnel found it in the idea of "partial dragging of the ether". According to this idea, the ether within bodies (e.g. within the glass of which the lens is made) is "dragged" by these bodies at a velocity $(1-1/n^2)\,V$, where V denotes the velocity of the body with respect to the ether and n is the refractive index. We can show that from Fresnel's hypothesis it follows that all the light propagation phenomena do not depend on the motion of the medium with an accuracy up to terms of the order of V/c. The possible differences are of the order of V^2/c^2; for the motion of the Earth around the Sun this ratio is about 10^{-8}. In 1881, A. A. Michelson carried out, for the first time, measurements intended to identify the Earth's motion with respect to the ether, taking into account effects of the order of V^2/c^2. Later, Michelson and Morley repeated the measurements. The latter, as we shall describe in greater detail in Chapter 4, gave negative results: it proved impossible to determine the Earth's motion with respect to the ether [34].

At the end of the 19th century, purely theoretical work was also done on the problem of light propagation and electrodynamics in moving media. It was then known that the wave equation

$$\frac{\partial^2\varphi}{\partial x^2}-\frac{1}{c^2}\frac{\partial^2\varphi}{\partial t^2}=0$$

is invariant with respect to Galilean transformations. Voigt [56] was the first to note in 1887 that this equation was invariant with respect to the transformation

$$x' = x-Vt,$$
$$t' = t-Vx/c^2,$$

containing the "local time" t, but without the relativistic square root $\sqrt{1-V^2/c^2}$. The latter appeared in 1892 in connection with a hypothesis proposed at that time by Lorentz [33] and Fitzgerald. According to their hypothesis bodies moving at a velocity V with respect to the ether were contracted in the direction of motion in the ratio of $\sqrt{1-V^2/c^2}:1$. This hypothesis explained the results of interferometric measurements by Michelson and Morley. In 1900, Larmor [33] gave formulae which are called Lorentz transformations while Lorentz himself demonstrated in 1903 the invariance of Maxwell's vacuum equations with respect to these transformations.

H. Poincaré was a true precursor of Einstein's theory. As early as 1895, he anticipated the relativity principle by writing that "experiments yield many facts which may be generalized in the following way: we cannot identify the absolute motion of matter with respect to the ether. We can only observe the motion of ponderable matter with respect to ponderable matter" [43]. In his later works, Poincaré criticized the notion of absolute time, formulated more precisely the relativity principle and tried to modify the law of universal attraction so that it would agree with the principle of the finite speed of propagation of interactions.

The 1905 paper by Einstein (*Zur Elektrodynamik bewegter Körper*, Ann. der Physik, **17**, 891 (1905)), who wrote it without knowledge of the 1903 work by Lorentz and the results of the experiments by Michelson and Morley, contained a novel, profound formulation of the relativity principle concerning electromagnetic phenomena. This paper has correctly been considered a turning point in the development of the special theory of relativity. Einstein's significant contribution was to complement the relativity principle with the principle of the independence of the velocity of light on the motion of the source. It turned out that the combination of these two simple principles led to a revision of the notion of time. From these principles Einstein was able to derive Lorentz's transformations directly; his predecessors had obtained them by considering transformations which did not change the form of Maxwell's equations. Einstein was the first to foresee the time dilation. In another paper in 1905, Einstein gave the relation between mass and energy, which was later popularized as the formula $E = mc^2$.

Minkowski's work [36], containing a geometrical, four-dimensional formulation of time and space and of Lorentz transformations, played an important role in the development of relativity theory. Shortly after having formulated the special theory, Einstein became interested in the problem of building a relativistic theory of gravitational phenomena. As early as 1907, observing that a uniform gravitational field is equivalent to the forces appearing in a system in uniformly accelerated, translatory motion, Einstein proposed that the relativity principle be extended to noninertial reference systems. Generalizing this observation and formulating the principle of equivalence between gravitational fields and inertial forces, Einstein predicted two phenomena: a change in the wavelength and the deflection of rays of light propagating in a gravitational field. Hence, it followed immediately that the special relativity theory could be valid only in the absence of gravitation. The phenomenon of the deflection of light rays suggested that the speed of light depended on the distribution of masses. Einstein's first theory of gravitation (1912) [11], also developed by Abraham [1], assumed that the speed of light was a position-dependent "gravitational potential" satisfying an appropriate field equation. At the same time, Nordström [36] proposed a relativistic theory of gravitation, also based on a scalar potential, but without the phenomenon of light deflection. Despite the fact that this phenomenon had not yet been confirmed, Einstein did not recognize any of those theories as satisfactory, because they did not treat on the same footing all the noninertial reference systems: they were not generally covariant. It became possible to satisfy the postulate of general covariance by introducing in 1913 the full metric tensor as a gravitational potential. Subsequent years were devoted to the search for the gravitational field equations; we shall say more about that in Chapter 11. In 1916 there appeared the paper [16] containing the final formulation of the Einsteinian, relativistic theory of gravitation. The following years brought its further development and confirmation. Already in 1916, there appeared Einstein's paper [17] devoted to an approximate method of solving the field equations and to gravitational waves, while Schwarzschild [48] found an exact solution of the field equations which now bears his name. Much attention and work was then devoted to the problems of gravitational energy and general covariance. In 1917 Einstein published a paper on cosmology [18], where he complemented the field equations with a "cosmological term", describing the hypothetical forces, sometimes presumed to occur, between distant astronomical objects. The modified field equations admitted a static model of the Universe, where the space (t = const) was a three-dimensional sphere with a radius proportional to the total mass of the Universe. In the 1920's, the expansion of the Universe (recession of the galaxies) was discovered, and in 1922 Friedmann [22] found a solution to Einstein's equations for the Universe involving expansion, without having to introduce the cosmological term.

Beginning with 1921, Einstein paid much attention to building a unified, geometrical theory of gravitation and electromagnetism. Research on this subject, initiated by Weyl [59] and Kaluza [30], did not achieve its aim, but it contributed to an understanding of the mathematical structure of field theory and to the development of the idea of geometrization of physics in connection with the theory of gauge fields.

The paper by Einstein and Grommer [19] which appeared in 1927 was devoted to the problem of the motion of bodies in the theory of gravitation. It turned out that, in contradistinction to other theories, in the relativistic theory of gravitation one could not postulate independently the field equations and the equations of motion of its sources; the latter were consequences of the former. The subject of the equations of motion was studied later again by Einstein, Infeld and Hoffmann. The fundamental work by these authors [21] contained the formulation of a new method for approximate and gradual solution of the field equations and for finding the motion. It is now known under the name of the EIH method. Independetly of Einstein, a similar method was developed by V. A. Fock and his students. Infeld and his collaborators carried out extensive research on the problem of motion in general relativity theory. It was summed up in a monograph [27] by Infeld and Plebański.

In the middle of the 1930's Einstein and Rosen attempted to find exact wave solutions of the gravitational field equations [20]. They both concluded that plane waves with a finite amplitude led to singularities in the geometry of spacetime. They found, however, physically satisfactory cylindrical waves. The paper by Einstein and Rosen initiated the development of extensive research on exact solutions of the gravitational field equations and the geometry of gravitational waves. Bondi, Pirani and Robinson [3], and other researchers, independently of the former, discovered the existence of exact, nonsingular plane gravitational waves. It appeared that the singularities encountered by Einstein and Rosen were in fact caused by the choice of the coordinate system and had thus no geometrical significance; R. Penrose explained the physical reasons for the occurrence of these apparent singularities. They were related to the fact that a plane wave acts as an ideal astigmatic lens. Light rays and particles going through such a lens are deflected. Particles moving parallel to each other before they come across the wave acquire convergent or divergent trajectories after crossing the wave; their world lines can intersect. This prevents the construction of the coordinate system proposed by Einstein and Rosen throughout the spacetime containing a plane gravitational wave.

By generalizing the properties of plane waves and using the methods developed in the course of analyzing them, it was possible to find large classes of solutions of Einstein's equations. It appeared that plane waves were important examples of so-called algebraically special solutions, according to the classification of gravitational fields initiated by Petrov [41]. An analysis of algebraically special fields led to the discovery of simple outgoing waves and of the Kerr metric describing the gravitational field and the geometry outside a rotating "black hole".

The middle 1950's witnessed the beginning of an intensive development of theoretical and experimental research in the field of gravitation. On the theoretical side, we should mention the numerous attempts to "quantize" the gravitational field, i.e. to build a microscopic theory of gravitational-interactions, by analogy with quantum electrodynamics. It seems that this problem has not been solved yet and that it is one of the most difficult problems of theoretical physics. Much attention has been paid to waves and gravitational radiation; numerous approximate methods have been elaborated for calculating the magnitude of radiated energy and of the effect of gravitational waves on the motion of bodies. Of particular interest is the research done on "black holes", their thermodynamical properties, and the phenomenon of creation of particle pairs by the strong gravitational field near black holes (the Hawking effect). Due to the studies of Hawking, Geroch and Penrose [24], it became clear that the reason for singularities occurring in the Friedmann cosmological models resided deep in the structure of Einstein's theory, rather than, as had previously been believed, in the high symmetry of these models. Papers published on the subject by E. M. Lifshitz, I. M. Khalatnikov and V. A. Belinsky played a major role. In connection with the problem of singularities, and also under the influence of the development of geometrical methods and gauge field theory, physicists have recently begun to analyze various versions of the relativistic theory of gravitation,; most of them are slight modifications of the Einstein theory. For example, the Einstein–Cartan theory allows torsion in spacetime and connects it with the spin of matter. In theories initiated by Chen Ning Yang, one considers Lagrange functions which are quadratic functions of curvature and torsion, in contrast to the Einstein theory, based on a linear Lagrange function. The boldest modification of Einstein's theory was the recently proposed theory of "supergravitation", where to describe gravitational interactions, physicists introduced an additional anticommuting field of particles with mass 0 and spin 3/2.

Of the experimental and observational works of the last two decades, it is interesting to note the measurements of changes in the length of electromagnetic waves in the Earth's gravitational field (Pound and Rebka, Jr. [44]) and the discovery of residual microwave radiation at a temperature of 2.7 K. (Penzias and Wilson [40], P. H. Dicke). This radiation, interpreted as a remnant of the hot development period of the Universe, is emphatic evidence for the Friedmann cosmological models. The discovery of residual radiation contributed to the fact that most experts abandoned the static model of the Universe, a model which enjoyed large popularity in the 1950's.

The development of radioastronomy and the technology for investigating outer space also provided other tests for the general theory of relativity. It proved possible to measure quite accurately the deflection and delay of electromagnetic waves in the Sun's gravitational field. The results of these measurements agree with the predictions of the Einstein theory, and they refute the Brans and Dicke theory, assuming that the general gravitational field should be described by means of a metric tensor complemented with a scalar field related to the gravitational "constant", which in this theory has no constant value. A count of distant radiogalaxies confirms Friedmann evolutionary cosmological models. We can draw a similar conclusion from an analysis of helium content in the Universe.

Much effort has been put into attempts to discover gravitational waves which are probably incident on the Earth from space. Particular hopes were connected with research by Weber [57], who initiated experimental studies in the field of gravitational radiation, building the first "gravitational antenna" in the form of an aluminium cylinder, equipped with sensitive piezoelectric detectors. Observations repeated by a few independent research teams have not confirmed Weber's first, opitimistic findings. The sensitivity of the antennae and detectors used until now is not sufficient to detect the gravitational radiation generated by double stars and supernova explosions in adjacent galaxies. Beginning his experiments, Weber hoped that previously unknown sources of gravitational radiation existed, with much greater intensities. It turned out, however, that it is necessary to improve the sensitivity of gravitational antennae by a few orders of magnitude in order to be able to detect on Earth the gravitational radiation from outer space. Current research in this direction is based on the use of antennae in the form of crystals with high quality factors and low temperatures, so as to eliminate the interference of thermal noise.

CHAPTER 2

Physical Phenomena, Models, Theories

We shall now consider the relation of the theory of relativity to other physical theories. Any discussion of the interrelationship between the different branches of physics is complicated by the fact that the way the discipline has been "pigeon-holed" in the course of its historical development does not fully agree with the present state of knowledge. Traditionally, we deal with areas such as mechanics, thermodynamics, electromagnetism, optics, elementary particle physics, solid state physics, theory of relativity, etc. Physicists are perfectly aware of the fact that these theories are only seemingly independent and that the division is inadequate.

To think sensibly about relationships between different physical theories and about a more appropriate, deeper division of physics, one has to get to the bottom of things and ask about the subject of physical studies. Without pretence to ostensibly learned definitions, it can be said that physics selects for study those natural phenomena which are relatively simple and for which it is possible to distinguish a few basic, characteristic quantities and name a few controlling factors. A no less important criterion is the repeatability of a phenomenon: physicists only study repeatable phenomena, which can be realized arbitrarily many times.

Because physicists restrict themselves to simple phenomena, they can model them. It is this modelling that is the basic element of the physicist's cognitive endeavour. Realizing this fact is essential to understanding what physical theories are and what are relations between them. We shall illustrate the phenomenon-model relationship on a few examples [28].

Consider the flight of a cannon-ball or, in other words, the motion of a projectile near the Earth's surface. In this phenomenon (in the common sense of the word) we can distinguish at least three physical phenomena, or groups of such:

(1) the motion of a body in a gravitational field;

(2) aerodynamic phenomena, including acoustic phenomena;

(3) thermal phenomena.

We shall concentrate solely on the mechanical aspects of the problem, i.e.

on the projectile motion in the narrow sense. As is well known, two factors controlling the motion can then be distinguished, namely, the initial velocity and projection angle, and gravity. If we neglect air-resistance, which can be done within a certain range of velocity and under certain assumptions concerning the mass, shape and size of the projectile, then neither the mass, shape nor size will have any effect on the motion. In general, finding what is significant in a given phenomenon is not easy and requires ingenuity and intuition on the part of the physicist. After the factors controlling the motion are asserted, it is possible, e.g. by photographic methods, to investigate the path of the cannon-ball and its progress in time (Fig. 2.1). Next, a mathematical law is

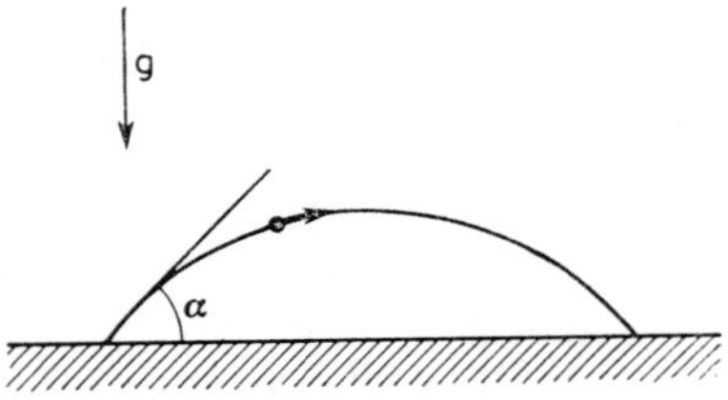

Fig. 2.1

developed to serve as the basis of a mathematical model of the phenomenon. As we know, the law has the form $m\ddot{\mathbf{r}} = m\mathbf{g}$, and all that can be said about the motion of the projectile is contained in this differential equation. If we wanted to include the effect of friction, all we would have to do would be to change the form of the equation a little, namely, to add a friction term to its right-hand side. Of course, before the equation is set down, we have to state that we are dealing with Euclidean space, in which the curves satisfying the equation are to be considered. The curves together with their parametrization describe both the path and the progress in time of the projectile. In this way we have a mathematical model of the phenomenon of projectile motion.

The next step was made by Newton, who discovered that the falling of bodies to Earth has the same cause as the motion of planets about the Sun or that of the Moon about the Earth. He then formulated a mathematical law which described all those phenomena: the law of universal gravity. Since it describes a wide class of phenomena, we say that we are dealing with a theory. The dividing line between a theory and a model is not clearly marked. Generally speaking, a model applies to a single phenomenon or to a group of similar phenomena, while a theory usually provides models for a wide range of phenomena, often seemingly disparate.

Our next, a little less trivial, example will concern the hydrogen atom. Historically, the first model of a hydrogen atom was that of Thomson, who represented the atom as a sphere of positively charged fluid, with a negatively charged point electron moving within it. The second was Bohr's planetary model. In this, we have a positively charged point nucleus and an electron moving about it in an ellipse according to the laws of classical mechanics (Fig. 2.2). To these laws, which follow from the principles of classical mechanics, Bohr added the so-called quantum conditions, which say that the momentum of the electron, and its action, may only assume certain defined values.

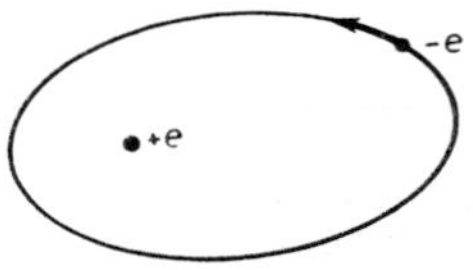

Fig. 2.2

Schrödinger's model, developed within the framework of a more general physical theory—quantum mechanics, came next. This model is entirely different from the previous two. The fundamental notion in it is that of Hilbert space, in which a certain Hermitian operator (Hamiltonian) is distinguished. The eigenvalues of this operator define the energy states of the hydrogen atom, and consequently the frequency of the light emitted by the atom. The eigenvectors of the Hamiltonian also have a physical interpretation: they provide complete information about the motion of the electron.

Historically, the next was Dirac's model, which drew on the concepts of relativistic quantum mechanics. In fact, it differs little from Schrödinger's model. There is some difference in the Hilbert space, and the form of the Hermitian operator is in accordance with special relativity theory.

In this way we have four (although there are more than that) mathematical models of the phenomena that make up a hydrogen atom, or—in short—four models of the hydrogen atom. The first two models are demonstrable, which cannot be said about the other two. By "demonstrable" we mean that the material points—and hence abstract mathematical objects—that we talk about in the first two models correspond to physical objects: a nucleus and an electron, and imagining this presents litle difficulty. Indeed, we go as far as to draw graphs of the functions obtained by solving equations of these models and consider them to be the pictures of the hydrogen atom. The situation in the Hilbert-space based quantum models is entirely different: they are subject to the uncertainty principle, which does not allow a definite position and momentum to be associated simultaneously with a given particle.

Quantum mechanics caused a much greater upheaval in the structure of physical theories than the theory of relativity. If a relativistic, but non-quantum, model of an atom were created, it would be as demonstrable as Bohr's model.

Dirac's model, on the other hand, is as non-demonstrable as Schrödinger's. The loss of demonstrability occurs in passing from non-quantum to quantum models, while no such loss exists between non-relativistic and relativistic models. In building a model of a physical phenomenon, we have to determine the following correspondence:

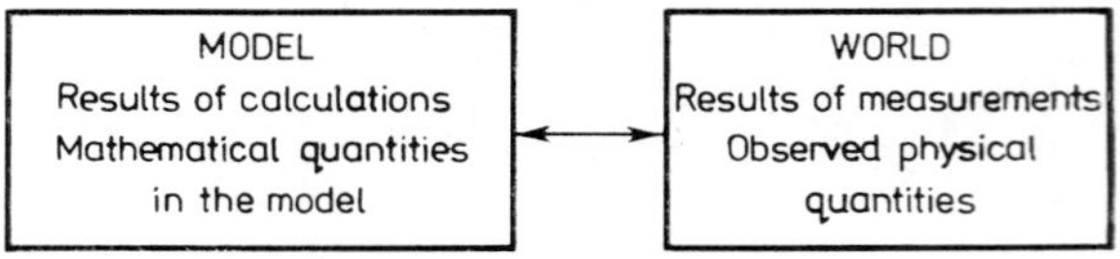

We must be able to answer the following question: which of the mathematical quantities that appear in the model correspond to the observed physical quantities, and more precisely, which of the calculated values correspond to the various measured values, and in what way? When we deal with classical models, this is quite easy.

In the case of projectile motion, for example, we have the differential equation $\ddot{\mathbf{r}} = \mathbf{g}$, which can be solved subject to some defined initial conditions. As a result we obtain certain functions of the variable t. Now if we want to give the position of the cannon-ball at some instant of time, we need to find the values of these functions for an appropriate value of the parameter t, which is interpreted as time. These values will determine the position (distance) of the ball with respect to a suitable frame of reference. The way the solutions of our differential equation should be related to the measurements of the actual position by means of measures, theodolites, photographic pictures, etc., can be prescribed very precisely. The same can be done for quantum models, although, owing to the loss of demonstrability, the problem is not as simple as before. By applying certain mathematical operations to the elements of the Hilbert space, we obtain numbers which can be compared with the results of the experiment. In this case, however, the prescription is not as obvious as in classical mechanics. Likewise, in the theory of relativity, the correspondence between the mathematical objects and that which is measured is not quite straightforward. Not everyone is aware of this, and a good many misunderstandings result. This remark applies to the "twin paradox" (the clock paradox), among others.

Having spoken of physical phenomena, of models and theories, we can give a diagram illustrating the cognitive method of physics:

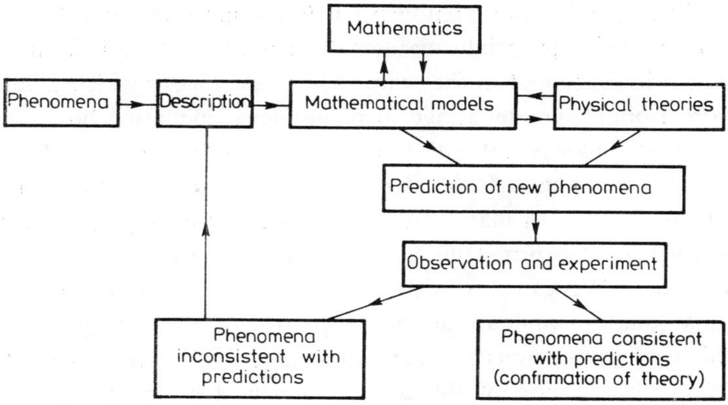

Experiment plays a fundamental role in the development of physics. In experimental work, physicists seldom follow any specific models; however, they use them for guidance in their choice and planning of experiments and observations. The description of a physical phenomenon, involving both qualitative and quantitative elements, can be treated as the first step towards its explanation. Then, by analysing the description, we construct a mathematical model. Usually such a model describes that single phenomenon or a group of similar phenomena. The next step is the development of a physical theory.

The task of a physical theory is to supply models for a wide class of phenomena; often these phenomena will appear to have very litle in common. We mentioned previously the theory of universal gravity, which provides models for both the projectile motion near the Earth's surface and the motion of planets about the Sun.

Usually it so happens that a theory permits us to describe more phenomena than were taken into account when it was developed. Whether or not new phenomena can be described by means of a given theory is often a criterion for accepting or rejecting it. A good example is furnished by the so-called older quantum theory, which was developed on the basis of Bohr's model of the hydrogen atom and which could only explain a very limited range of atom-related phenomena, and even these required additional assumptions. In the sense of what has been said so far, it is even difficult to speak of the older quantum theory as of a theory; it should rather be looked upon as a few models having certain properties in common.

Returning to Newton's theory of gravity, it is worth recalling that it produced models of phenomena that had not been considered at the time it was created. For example, it was capable of explaining ocean tides. If, basing ourselves on a theory, we predict a phenomenon, then we have to get back to the starting point, that is to experiment, to find out whether the model put forward by the theory describes that phenomenon satisfactorily. If not, we should analyse in detail the phenomenon that is inconsistent with the theory and try to give an appropriate mathematical model for it. This model must then be taken into account in the search for a new theory which would be capable of describing a greater range of phenomena, including both those consistent and those inconsistent with the predictions of the original theory.

In what sense do models of physical phenomena have the character of mathematical models? In many cases—literally. For example, the differential equation which models projectile motion is a fragment of the theory of differential equations. We can say with full precision all about the methods of solving this equation and about the properties of its solution. If we leave aside (which we should not) the very important part of any physical theory, namely the determination of the correspondence between the mathematical quantities and the results of measurements, then many physical theories will be reduced to a piece of mathematics. Thus, practically the whole of classical mechanics becomes a chapter of the theory of ordinary differential equations. The situation is similar with classical electrodynamics: if the existence of point charges is neglected and only continuously distributed matter is considered, then the entire Maxwell theory becomes a section of the theory of partial differential equations.

Often, however, physicists do not restrict models of physical phenomena to the existing mathematics but create something which they hope may eventually become a mathematical model. An example of this is Dirac's δ function. It was introduced by physicists even though everybody with any knowledge of the theory of integral could easily prove that such functions did not exist. Physicists used that not very precise concept for more than twenty years before mathematicians created the theory of distributions, which explained what really those "functions" meant, thereby sanctioning their use.

Another, more recent example are the generalized Hilbert spaces. It turned out that in some problems of quantum mechanics ordinary Hilbert spaces were no longer sufficient as, for example, it proved convenient to speak of vectors of infinite length even though no such vectors existed in Hilbert spaces. Today, no contradiction exists here either, for appropriate mathematics has been developed which legitimizes the use of these concepts (Gelfand's triplets, bristling Hilbert spaces).

There are still models at the present time, or should we more accurately call them outlines of models, the formulation of which as strict mathematical models remains an open question. We have in mind quantum field theory in the form in which physicists use it for interacting fields. Certain assumptions which are made in this model are known to be contradictory. Quantum field theory produces results comparable with observations by using approximate calculations in which an essential part is played by renormalization. This process involves eliminating divergent quantities ("infinities") that appear in every such theory, for example in quantum electrodynamics. It is not clear whether this outline of a model, known as quantum field theory, can ever be presented in a completely satisfactory logical form, as was possible with Dirac's δ function. This question is at present among the most important problems of theoretical physics.

Physicists hardly ever mention the fact that most of their work is concerned with models. Usually, we say: "consider a hydrogen atom according to quantum mechanics", which suggests that we are really considering an atom of hydrogen. What we do consider in fact is a mathematical model of a hydrogen atom proposed by quantum mechanics, hence a creation of human mind. Why does not anybody put it this way? Firstly, because leaving out the expression "mathematical model" is convenient and, although imprecise, in general it does not lead to mistakes. Secondly, it is more pleasant to entertain the illusion that we are dealing directly with the objective reality and not only its mathematical models. Physicists, of course, are full of best intentions and want to speak about that reality, but the only reasonable way, it seems, is to construct models.

It is worth realizing, however, that theoretical physicists deal primarily with models. Sometimes this fact does lead to misunderstandings. For example, one often hears the question: does gravitational radiation exist? An unreasonable answer is sometimes given: why, Einstein has long proved that, so why should physicists keep on trying to detect this radiation experimentally. Such an answer is, of course, a misapprehension. Einstein investigated gravitational radiation within a model proposed by his own theory. Certain solutions of the equations of gravitational field turned out to be similar to some of the solutions of Maxwell's equations, which, as is well known from experiments, describe electromagnetic waves. There is no certainty, however, whether Einstein's model of gravitational field is fully adequate, and particularly, whether it is true in the part dealing with gravitational radiation. Naturally, if a theory describes many different phenomena well, we tend to believe that other predictions of that theory will also be confirmed experimentally; but this does not have to be so.

It now becomes clear what is meant by saying that a physical theory can be admitted to be true or false. Sometimes we read or hear a statement that someone refuted a theory, for example that Einstein refuted Newton's theory. If we accept that the main substance of a physical theory are models which describe the reality only approximately, then the above statement does not make much sense. At best, we can produce new models which will describe the phenomena concerned better than existing models. This is exactly the case with Einstein's theory: models proposed by it were more accurate than those proposed by Newton, but saying that Einstein refuted Newton's theory is not correct. Newtonian models continue to be applied and there is no indication that they may cease to be applied. This is so because they give good description of phenomena within a certain range of velocities, namely under velocities small compared with the velocity of light.

It sometimes happens that there are two competing models aspiring to describe the same range of phenomena and in the same approximation, and one of them appears to give results consistent with observations while the other does not. Then, we are inclined to state that the first model is good and the second is bad, and perhaps even reject the second model out of hand. Cases like this, however, are rather rare. Much more often we have the situation where a number of models fit the reality, but each for a somewhat different range of phenomena. We may then try to improve on these models, i.e. to replace them by new models which have a wider range of applicability.

There is hardly a physicist nowadays who considers the existing theories to be final. This statement applies particularly to quantum mechanics and the theory of relativity. No one now thinks that these theories are the last word. It seems certain, however, that there is no way back. Everybody is convinced that physics will not develop in a direction where we would have to discard quantum mechanics and turn towards classical models. This is impossible simply because classical models have a limited range of applicability in comparison with quantum models. A similar remark can be made about the theory of relativity. But then nobody says that this theory will never be improved or replaced by a broader, more exact, more general theory.

In what other, more correct, way could we divide physics? The division can be made with regard to different aspects of the investigated phenomena. The first division that comes to mind is one with respect to the fundamental interactions that occur in the physical phenomena. At present we distinguish the following kinds of actions, or forces:

gravitational,
electromagnetic,
nuclear $\begin{cases}\text{weak,}\\ \text{strong.}\end{cases}$

We have listed these forces in the order in which they were discovered. Gravitational and electromagnetic forces are believed by the majority of physicists to have been well investigated and understood. Our knowledge of nuclear forces is still far from adequate. There is no satisfactory theory capable of describing a sufficiently wide range of phenomena involving these forces. Indeed, it is possible that the above classification of nuclear forces is not appropriate.

Naturally, it almost never happens that forces of only one kind are present in a phenomenon. It is usually possible, however, to distinguish one dominant type and often all other interactions can then be disregarded. For example, in considering the motion of the planets about the Sun we find that it is dominated by gravitational phenomena and that the contribution to it from all other forces is negligible. Likewise, where the structure of an atom is concerned, the dominant part is played by electromagnetic forces; gravitational interactions between the nucleus and the electrons are some 10^{40} times weaker than the electromagnetic ones.

The second division of phenomena refers to their scale, to the number of elementary components involved. We can consider phenomena on a scale of the micro-world, i.e. that of single elementary particles: photons, electrons, nucleons and all others that have been discovered in the last 40 years. We can study atomic nuclei, atoms, ions and molecules. We can also consider matter on a macroscale, e.g. this book as a material object, which consists approximately of 10^{24} different molecules. Then we can look at matter on an astronomical scale, thus at planets, stars, stellar constellations. Finally, we can ponder on the Universe as a whole. Appropriate to each of these scales, separate theories are developed.

It is generally believed that the theories relating to the phenomena of the microworld, and hence those describing single elementary systems, are the most fundamental. It is thought that it should be possible to derive the theories describing macroscopic phenomena from the microscopic theory by allowing for the large number of particles that take part in those phenomena.

The third division, rather less significant than the previous two, concerns the velocity of the motions involved. It appears that the velocity of light, c, is the critical one, with which all other velocities should be compared. Phenomena involving motions with velocities much lower than the velocity of light can be described within the framework of Newtonian models, while those with velocities significant compared to c require the use of the theory of relativity. Relativistic descriptions of low-velocity phenomena differ very little from Newtonian descriptions, which is why Newton's theory is often spoken of as a limiting case of the theory of relativity.

Finally the fourth division, related to the second, is concerned with the applicability of classical and quantum methods. In practically all situations in which we deal with single objects of the microworld we have to use the laws of quantum mechanics. When we describe systems composed of a large number of such objects, we can in most cases apply classical theories. Quantum laws are more general; by a suitable limiting process we obtain from them the classical laws.

The above divisions are not quite independent. For example, the phenomenon of electromagnetic radiation is by its very nature relativistic, because electromagnetic disturbances propagate at the velocity of light. It follows that the most exact theory of phenomena involving individual photons is both quantum-based and relativistic. It is quantum electrodynamics. Developed in the 1930's this theory now offers the most precise quantitative verifications of the theory of relativity. Indeed, a frequent error of those who try to challenge the theory of relativity is that they forget how rich and precise the predictions are of relativistic quantum electrodynamics.

Consider as an example one of the results of quantum electrodynamics: the formula which gives the energy states of the hydrogen atom.

$$E_{njl} = -\frac{me^4}{2\hbar^2 n^2}\left(1+\frac{\alpha^2}{2n^2}\,\frac{4n-3j-3/2}{2j+1}-\frac{8\alpha^3}{3\pi n}Q_{njl}+\ldots\right),$$

$$Q_{njl} = \begin{cases} \ln\dfrac{mc^2}{2\varepsilon_{nl}}+\dfrac{19}{30}, & l=0,\\[2ex] \ln\dfrac{me^4}{2\hbar^2\varepsilon_{nl}}+\dfrac{3}{8}\,\dfrac{C_{jl}}{2l+1}, & l\neq 0,\end{cases}$$

$$C_{jl} = \begin{cases} \dfrac{1}{l+1}, & j = l+\dfrac{1}{2},\\[2ex] -\dfrac{1}{l}, & j = l-\dfrac{1}{2}.\end{cases}$$

The following notation is used:

m—electron mass,
e—electron charge,
$\hbar$—Planck constant,
$\alpha = e^2/\hbar c$—fine structure constant ($\alpha \cong 1/137$),
ε_{nl}—mean excitation energy of state nl.

The indices n, j and l number the different states; n and l take integer values, $n = 1, 2, 3\ldots$, $l = -n, -n+1, \ldots, n-l, n$; for a given l, the index j assumes one of the two values $j = l+1/2$ or $j = l-1/2$. The magnitude of the successive terms inside the brackets depends above all on the power in which the fine-structure constant α occurs. The further, dotted terms, which involve higher powers of α, are much smaller. The second term in brackets was obtained in Dirac's theory, the third in quantum electrodynamics, and the subsequent dotted terms in the relativistic versions of both theories. Thus every new quantum theory contributed a correction to the previous one. The formula has proved to be in a very good agreement with experiment. Discrepancies are of the order of 10^{-10}. This is of course the great success of this theory.

Today quantum electrodynamics is considered to be the most precise of physical theories. It should be borne in mind that it rests on three pillars. The first is Maxwell's electromagnetic theory, the second—quantum theory, and the third—the theory of relativity. If any one of them is removed, the whole theory, so admirably concordant with experiment, will simply collapse.

Every now and then a paper appears whose author proposes a theory that gives an explanation different from Einstein's of the Michelson–Morley experiment and other simple experiments. On this basis that author then claims that the theory of relativity can be replaced by his own theory. In fact, there are many theories explaining the Michelson–Morley and a few other experiments, but one must not forget that there are a great many experiments whose results agree very precisely with the theory of relativity. It is worth remembering that, as in medicine, the standing rule here is: above all, cause no harm. If one wants to replace a theory of relativity by something new, then the new theory should predict the results of the same experiments with at least the same precision as the theory of relativity. Quantitative results of a new theory should agree with measurements to at least as good an accuracy as those produced by the old theory.

It is known that a theory that would account for all facts concerning elementary particles does not exist. Some partial models, however, have been successfully developed. This applies, for example, to reactions between elementary particles. Physicists have no ready model which would be good enough to describe these reactions; the kinematics, however, i.e. that part of the theory which does not require for description the forces causing the reactions, has been well worked out and is consistent with experiment. It is based on the theory of relativity, among other things. It is certain that this partial model will go to the making of a future "good" theory of elementary particles.

CHAPTER 3

Galilean Spacetime

In analysing physical phenomena physicists separate out several different aspects, which are the subjects of study of different branches of physics. We illustrated this in the preceding chapter in the example of the flight of a cannon-ball. There are certain aspects which are common to all natural phenomena, namely those concerning the time and space relationships between the material objects involved. Hence underlying almost every model of physical phenomena is a model of time and space relationships, or spacetime. It is built by abstraction from "what happens" to only the "when and where". We can consider phenomena that occur close to each other, in a limited region of space and time, or in the Universe as a whole. Thus we may need to consider different models of spacetime.

Before the general theory of relativity was formulated, the prevalent view was that the properties of space and time were invariant and everywhere the same. It was thought that the local properties of spacetime determined its global properties and that a model based on the observation of phenomena in our immediate neighbourhood applied to the Universe as a whole. Since Einstein's time we know that this is not so.

The spacetime of a nuclear physicist differs from that of a cosmologist because, for example, the latter must take into account the curvature of space, while it seems unnecessary to allow for it in atomic-scale phenomena and still less so in nuclear phenomena. Thus we speak not of one space but of many.

The notion of spacetime originated together with the theory of relativity. Before Einstein, time and space were spoken of separately. However, if we think about it, we come to see that something which should be called spacetime according to the present terminology was considered even before Einstein. We will soon see that, in spite of what is often said, time and space are not completely separated in Newtonian mechanics.

In the language of modern mathematics: Newtonian spacetime cannot be naturally represented as a Cartesian product of time and space.

What is spacetime? It is a set of elements called events. In keeping with what was said earlier, we must define the relationship between the concepts which occur in a model and that which we observe in reality. In particular, we must say what the mathematical concept of a point-event corresponds to.

We obtain it by abstraction from what is called an event in everyday language.

Let us turn back to the example of the cannon-ball. Consider the event of the ball hitting the ground. We note that it lasts a certain period of time and occupies a certain region in space. For our purposes, however, it is more convenient to assign to it a specific instant of time and a defined point in space. For example, we may choose the moment a selected point of the ball first touches the ground and the point at which this contact occurs. We then only need to forget, as inessential, what is actually happening, and we come to the concept of an event.

The same can be done with any other event that takes place during the flight of the cannon-ball, as is schematically shown in Fig. 3.1. We can see that

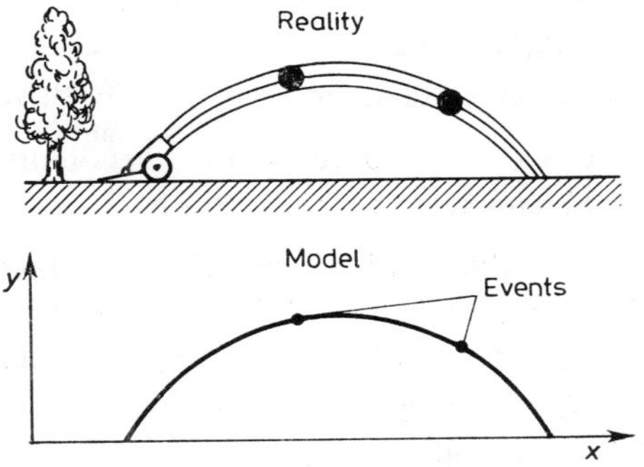

Fig. 3.1

the concept of an event is obtained by abstraction in three ways: firstly, we abstract from what really happened, secondly—from how long it took, and thirdly—from how much space it occupied.

Theories of time and space differ from one another in the mathematical structures which they adopt for spacetime. All serious theories, however, have one property in commom: they assume that spacetime is a four-dimensional differential manifold. To explain this concept, we give a few mathematical definitions.

Chart. Let E be a set (later E will denote spacetime) and A a subset of E: $A \subset E$. Let ξ be a map of A onto an open subset of n-dimensional arithmetic space $\mathbf{R}^n$ (Fig. 3.2). We require in addition that ξ be a one-to-one mapping.

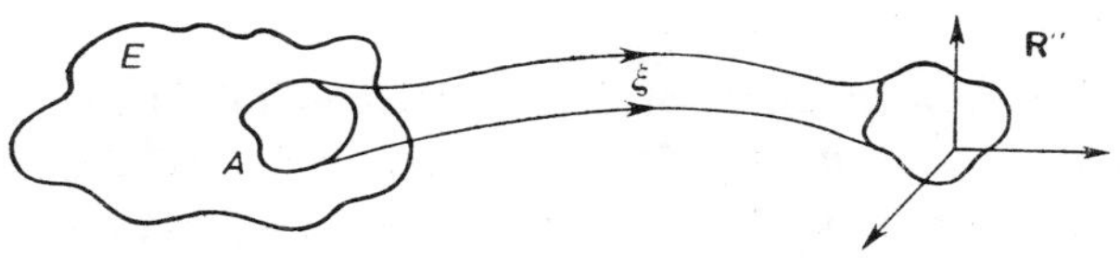

Fig. 3.2

The pair (A, ξ) is called a *chart*, A the domain of the chart, and ξ its coordinate system. The concept of a chart corresponds to what a chart is in geography. There the set E is the surface of the Earth, the subset A a portion of that surface, and the mapping ξ defines the way in which points of that portion are associated with points on a sheet of paper. This chart is of course two-dimensional.

Compatibility of charts. Let (A, ξ) and (B, η) be two charts on the set E. If $(A \cap B, \xi|_{A \cap B})$ and $(A \cap B, \eta|_{A \cap B})$ are charts (in other words, if $\xi(A \cap B)$ and $\eta(A \cap B)$ are open sets) and, furthermore, the maps $\xi \circ \eta^{-1}|_{\eta(A \cap B)}$ and $\eta \circ \xi^{-1}|_{\xi(A \cap B)}$ are k-times continuously differentiable, then the charts are said to be C^k-compatible. The number k may take value from 0 (we then speak of topological compatibility of charts) to ∞. We can also consider charts which are analytically compatible; we then write that they are C^ω-compatible.

In the language of geography, compatibility of two charts means that towns which lie close to one another on one chart must do so on the other, and that rivers with smooth sides on one chart must have smooth sides on the other.

Atlas. An atlas of class C^k is a collection of charts mutually C^k-compatible whose domains cover E. This term is also taken from geography.

Differential manifold. A differentiable manifold of class C^k ($k = 1, 2, \ldots, \infty$ or ω) is a set E with a maximal atlas of class C^k (i.e. such that no new charts can be added to it without violating compatibility). The dimension of the differentiable manifold E is defined as the dimension of the arithmetic space $\mathbf{R}^n$ on which the manifold is modelled.

Any atlas can be extended to a maximal atlas in a unique way. A manifold is usually assumed to have a countable atlas (as a rule, the maximal atlas is not countable). We will assume that spacetime is a differentiable manifold of class C^∞. In addition, we will assume that it satisfies the Hausdorff axiom, although some say that in strong gravitational fields this axiom may not hold.

The Hausdorff axiom: *For every pair of distinct points of the manifold E there exist two charts in the atlas of E such that their domains are disjoint and each contains one point of the pair.*

The reader may have noticed that the above definition of a differentiable manifold is somewhat different from that usually given in books on differential geometry: we did not assume in advance that the set E is a topological space. Of course both approaches are equivalent, since the topological structure can be introduced by requiring that E has the weakest topology in which the coordinate systems are still continuous.

We now give a few examples of differentiable manifolds.

1. Let E be an open subset of $\mathbf{R}^n$. The chart $\xi: E \to \mathbf{R}^n$ defined by the formula $\xi(x) = x$ is an atlas on E. The set E and this atlas constitute a differentiable manifold of class C^ω, i.e. an analytic manifold.

2. Consider the n-dimensional sphere $\mathbf{S}^n$, i.e. the set of points $(x_1, \ldots, x_n, x_{n+1}) \in \mathbf{R}^{n+1}$ satisfying the condition

$$x_1^2 + \ldots + x_n^2 + x_{n+1}^2 = 1.$$

An atlas consisting of only one chart cannot be defined on the sphere if we want to maintain its natural topological structure. This is because the sphere is a compact set and therefore cannot be mapped continuously onto an open subset of $\mathbf{R}^n$. We can, however, introduce an atlas consisting of two charts. Let A and B be the subsets of $\mathbf{S}^n$ obtained by removing from it the north and south poles, respectively, i.e.

$$A = \mathbf{S}^n - \{(0, \ldots, 0, 1)\},$$

$$B = \mathbf{S}^n - \{(0, \ldots, 0, -1)\}.$$

Define the coordinate system ξ: $A \to \mathbf{R}^n$ and η: $B \to \mathbf{R}^n$ by the formulae

$$\xi(x_1, \ldots, x_n, x_{n+1}) = \left(\frac{x_1}{1-x_{n+1}}, \ldots, \frac{x_n}{1-x_{n+1}}\right),$$

$$\eta(x_1, \ldots, x_n, x_{n+1}) = \left(\frac{x_1}{1+x_{n+1}}, \ldots, \frac{x_n}{1+x_{n+1}}\right).$$

These coordinate mappings are known as stereographic projections (cf. Chapter 7). The reader will readily verify that the mappings $\xi \circ \eta^{-1}$ and $\eta \circ \zeta^{-1}$ are analytic. Thus we can furnish the n-sphere with the structure of an analytic manifold.

3. If E_1 and E_2 are differentiable manifolds of classes C^{k_1} and C^{k_2}, respectively, then the Cartesian product $E_1 \times E_2$ is endowed with a natural differentiable structure of class C^k, where $k = \min(k_1, k_2)$. For, let (A_1, ξ_1) and (A_2, ξ_2) be two charts on E_1 and E_2, respectively: $A_1 \ni p_1 \mapsto \xi_1(p_1) \in \mathbf{R}^{n_1}$, $A_2 \ni p_2 \mapsto \xi_2(p_2) \in \mathbf{R}^{n_2}$. Define the mapping $\xi_1 \times \xi_2$ as

$$(\xi_1 \times \xi_2)(p_1, p_2) = (\xi_1(p_1), \xi_2(p_2)) \in \mathbf{R}^{n_1} \times \mathbf{R}^{n_2}.$$

Then the pair $(A_1 \times A_2, \xi_1 \times \xi_2)$ becomes a chart on $E_1 \times E_2$ and the collection of all charts of this form constitutes a C^k-atlas on $E_1 \times E_2$.

Using points 1 and 2 above, we can see, for example, that the cylinder $\mathbf{R} \times \mathbf{S}^1$ and the torus $\mathbf{S}^1 \times \mathbf{S}^1$ are manifolds of class C^ω.

4. A Lie group is an abstract group which is a differentiable manifold satisfying the Hausdorff axiom, and in which the group operations, i.e. multiplication and inverse, are continuous.

5. All the examples of differentiable manifolds given so far satisfied the Hausdorff axiom. There are manifolds, however, which do not satisfy it. As an example, consider the manifold consisting of two open-ended rays and two additional points, as in Fig. 3.3, with charts defined by the projections on

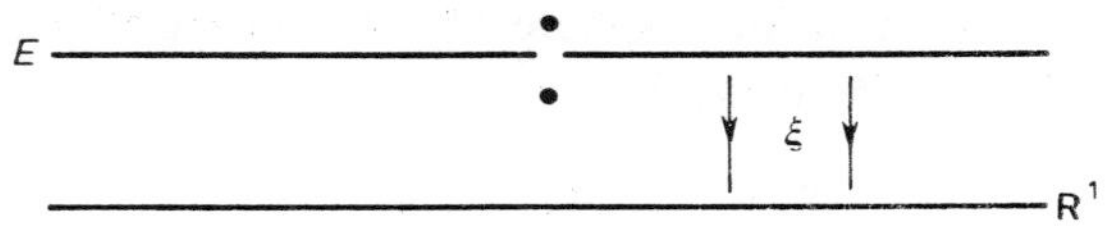

Fig. 3.3

a straight line representing $\mathbf{R}^1$. It is not difficult to verify that this indeed is a manifold and that it does not satisfy the Hausdorff axiom: no two charts will each contain exactly one of the two isolated points.

To sum up, we have assumed that spacetime is a four-dimensional differentiable manifold of class C^∞ with a countable atlas, satisfying the Hausdorff axiom. From now on, the symbol E will denote just such a spacetime.

Until recently, when speaking of reference systems, physicists used objects such as standard weights, rigid measuring rods, and clocks. In the last decade the situation changed, particularly where measuring rods are concerned. Physically, considering the atomic structure of matter, they are very complex systems. They are susceptible to many external influences: pressure, temperature, various fields; so that, in reality, they are never rigid. This was the reason for the "dethronement" of the famous platinum-iridium standard metre of Sèvres. Today, the formally accepted definition of a metre is based on the wavelength of a certain atomic radiation. This is because electromagnetic radiation has a relatively simple structure and is almost independent of the external conditions, which makes the constancy of its wavelength easy to maintain. Therefore, in the present book we will speak of light signals rather than measuring rods, both in the context relating to coordinate systems and in that of length measurements. In addition to light rays, we shall use ideal clocks, which can be realized as, say, "nuclear clocks"—practically insensitive to external influence.

How can we realize the coordinate systems that appear in the definition of a differentiable manifold? The mathematical concept of a coordinate system corresponds to that of a reference frame. Reference frames can be imagined as clocks, with which we fill in the entire space, each with three numbers engraved, defining the position (Fig. 3.4). The clocks need not be ideal, they do not even have to keep good time, as long as cloks situated close to one

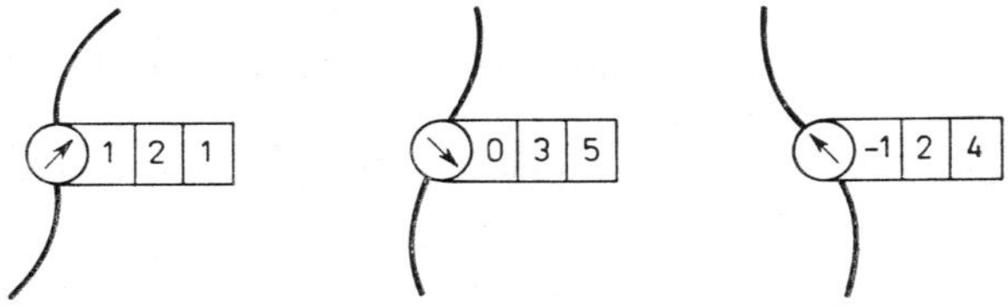

Fig. 3.4

another show similar times. In addition, the numbers engraved on neighbouring clocks should also be close.

Let us note that the reference frames so introduced have no metric properties. To introduce a metric structure, we must consider certain special physical phenomena such as the propagation of light and "good" clocks.

What is the simplest way in which to introduce the structure of a differential manifold without resorting to the concepts or objects that occur in metric problems, like measuring rods or—in our case—clocks, is a question worthy of study. More specifically, it would be interesting to know whether spacetime can be furnished with the structure of a differential manifold solely by means of light signals. This problem was studied by S. Woronowicz in his doctoral dissertation, in which he showed that simple observations of light signals are indeed sufficient to introduce the structure of a topological manifold. Attempts to introduce a differentiable structure on the same bases have so far failed.

Sometimes we hear the opinion that the idea of four-dimensionality is associated with the theory of relativity and that world used to be three-dimensional before this theory came into being. Of course this is not so. Spacetime was four-dimensional before Einstein too. To understand what this four-dimensionality means, consider how we start a letter: we specify the place where we are and give the date, or the time of the event of writing this letter. Here lie those four numbers: the altitude above the sea level (we often give it in postcards from mountainous regions), latitude and longitude (sailors write these), and time (can be represented by a single number, e.g. the number of seconds after the birth of Christ). Space is three-dimensional in the theory of relativity too, only the division of spacetime into space and time is different from that in Newtonian physics.

A differentiable manifold is a very general concept, and a spacetime assumed to be no more than just a differentiable manifold would be insufficient to describe most phenomena. For this reason, we have to introduce in our spacetime certain additional geometrical structures. One such structure, the affine connection, also called linear connection or parallel transport, is common to all theories of space and time. We shall discuss it later. For the present, let us consider those special models which in providing a basis for the description of phenomena disregard gravitational interactions.

To understand why we exclude gravity, one must appreciate the exceptional character of gravitational interactions. The most important feature distinguishing gravitation from all other forms of interactions is that the motion of a body in a gravitational field does not depend on the properties of that body but only on the properties of the field itself. We know, for example, that near the surface of the Earth all bodies fall with equal acceleration, customarily denoted by g. No such independence holds for other kinds of interactions.

In the case of an electromagnetic field, for example, there exist bodies electrically neutral, which are left unaffected by the field. Furthermore, we can completely eliminate the effect of the field by erecting a suitable screen. No such screen is possible against a gravitational field, and this is its second distinguishing feature. A more detailed analysis shows that gravitational interactions can be reduced (because of these special features) to certain properties of spacetime. However, this greatly complicates the geometry of spacetime, so that, for the present, we shall leave gravitation aside.

Under this assumption, we have the first law of dynamics, which says that

(1) there exists a preferred class of motions, called free motions,

(2) there exist reference frames relative to which the free motions have no acceleration.

Usually, point (1) is further clarified by saying that a body is in free motion when no external influences act upon it. Point (2) is usually formulated in a somewhat different manner, namely, that free motions are rectilinear and uniform relative to certain reference frames called *inertial frames.* The two formulations are shown to be equivalent if spacetime is furnished with a geometrical structure which admits the concepts of rectilinearity and uniformity. Such a structure is provided by the affine space.

Affine space. An affine space is a pair (E, V), where E is a set and V a vector space, with a mapping "+": $E \times V \to E$, such that V acts in E as an (Abelian) group of transformations, i.e. for arbitrary elements $p \in E$ and $u, v \in V$ we have

$$(p+u)+v = p+(u+v)$$

and, if 0 denotes the zero element of space V,

$$p+0 = p \qquad \text{for every} \qquad p \in E.$$

In addition, the action of V in E is required to be transitive and free. The freedom means that the equality $p+u = p$ should imply $u = 0$. Transitivity, on the other hand, requires that for any $p, q \in E$ there must exist $u \in V$ such that $p+u = q$. It is readily seen that, owing to the assumed freedom, if such a vector exists, it is unique; it is then called the difference between q and p, and written $u = q-p$.

Dimension. The dimension of an affine space (E, V) is the number equal to the dimension of the vector space V.

If V is a vector space, then (V, V), with the mapping "+" defined by the addition of vectors in V, is an example of an affine space. All affine spaces of equal dimension are isomorphic. If, in an affine space, we distinguish a point $\mathfrak{o} \in E$, then to every point $p \in E$ we can assign a unique vector $u(p)$ by the formula $p = \mathfrak{o}+u(p)$. The mapping $p \to u(p)$ realizes an isomorphism between the affine spaces (E, V) and (V, V). Although, as we can see, an affine space is isomorphic to a vector space, the two notions should not be identified. In an affine space no element is distinguished, while in a vector space there is one such element: the zero vector.

Repère. A repère or basis, of an affine space (E, V) is the pair $(\mathfrak{o}, e)$, where $\mathfrak{o} \in E$ and $e = (e_1, \ldots, e_n)$ is a basis of the n-dimensional vector space V.

Every point $p \in E$ can be uniquely represented in the form $p = \mathfrak{o} + u(p)$, where $u(p) \in V$. Then we can decompose the vector $u(p)$ with respect to the basis e, whereupon we obtain

$$p = \mathfrak{o} + \xi^i(p) e_i .$$

Notation in this formula follows Einstein's summation convention: if an index is repeated once at the lower level and once at the upper level, the summation must be carried out over the whole range of that index. The above formula defines a one-to-one mapping $\xi: E \to \mathbf{R}^n$, $\xi = (\xi^1, \xi^2, \ldots, \xi^n)$. We can see that the pair (E, ξ) is a chart, the domain of which is the whole space E. This single chart can be treated as an atlas on E, and then E becomes a differential manifold with the dimension equal to that of the affine space (E, V).

Affine transformation. An affine transformation of an affine space (E, V) is a pair of one-to-one mappings, (f, α_f), where $f: E \to E$ and $\alpha_f: V \to V$ is a linear transformation, such that

$$f(p+u) = f(p) + \alpha_f(u) .$$

As follows from this definition, α_f is a one-to-one and onto map, thus a bijection. Similarly, f is a bijection. The mapping α_f is uniquely determined by f, which is why an affine transformation is often identified with the mapping f alone.

Let us find the form of our affine transformation with respect to a certain fixed repère $(\mathfrak{o}, e)$. Let $p = \mathfrak{o} + \xi^i e_i$ and $f(p) = \mathfrak{o} + \xi'^i e_i$. We have

$$f(p) = f(\mathfrak{o} + \xi^i e_i) = f(\mathfrak{o}) + \xi^j \alpha_f(e_j) .$$

Next, $\alpha_f(e_j) = \alpha^i_j e_i$, where (α^i_j) is an invertible matrix. Also, the point $f(\mathfrak{o})$ can be written

$$f(\mathfrak{o}) = \mathfrak{o} + \big(f(\mathfrak{o}) - \mathfrak{o}\big) = \mathfrak{o} + \beta^i e_i .$$

Finally we obtain

$$\xi'^i = \alpha^i_j \xi^j + \beta^i ,$$

where $\det(\alpha^i_j) \neq 0$. An example of affine transformation is a translation $f(p) = p + u$, where u is a constant vector. In this case, the matrix (α^i_j) is the unit matrix, $\alpha^i_j = \delta^i_j$.

Since spacetime is a four-dimensional manifold, from now on we shall assume that the dimension of the affine space (E, V) (and hence that of the manifold E) is 4.

Straight line. A straight line in E is a set of points of form $\{\lambda u + p : \lambda \in \mathbf{R}^1\}$, where p is a fixed point in E and u a fixed vector in V (Fig. 3.5). The vector u

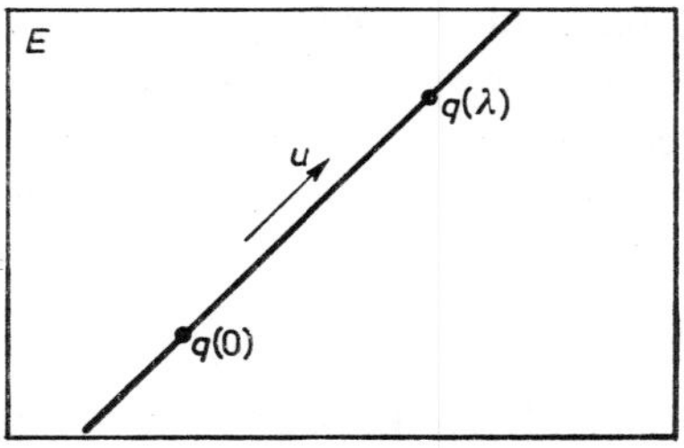

Fig. 3.5

is called the directional vector of the line; it is determined up to a scalar factor. The point p is not uniquely determined by the line either: any other point of the line can be taken instead. The parametric equation of the straight line has the form

$$q(\lambda) = u\lambda + q(0).$$

If we denote $x^i = \xi^i \circ q$, then

$$x^i(\lambda) = u^i\lambda + x^j(0),$$

where u^i are the coordinates of the vector u in the basis e, that is $u = u^i e_i$.

The coordinate system $\xi : E \to \mathbf{R}^4$ defined by an arbitrary repère $(\mathfrak{o}, e)$ is called a *rectilinear system*. The term derives from the fact that the curves determined by the requirement that three of the coordinates be constant are straight lines parallel to the basis vectors. For example, by holding ξ^1, ξ^2 and ξ^4 constant we obtain a straight line for which e_4 is a directional vector. Naturally, systems other than rectilinear can also be introduced; they will be considered in Chapter 10.

From its parametric equation we can see that a straight line is given by a linear relationship between the coordinates (after λ is eliminated). Remembering the 1st law of dynamics, we conclude that these are the straight lines which are the world-lines of free motions. By a *world-line of a material point* we understand the set of all events belonging to its history. The world-line of a material point, which is a curve in spacetime, must be distinguished from the trajectory of this point, which is a curve in three-dimensional space.

What will happen if we change the basis, $e \mapsto e'$? Clearly, the coordinates will undergo a linear transformation. Of these coordinates, one should be called time. As a result of the tansformation, we will obtain a new "time", which will be a linear combination of the old time and the old spatial coordinates. Such a linear transformation of time may seem to be unacceptable.

We will give an example, however, showing that similar (linear) transformations of time can be performed even in ordinary life.

As is well known, there exist time zones on the Earth. We can imagine an international agreement (putting aside its impracticality), according to which the time zones are replaced by a continuous distribution of time. Then, if we travelled 1° east-ward, we would have to advance our watches by 4 minutes.

Now take a not-too-large region on the Earth's surface and approximate it with a plane, as has been done in Fig. 3.6. Imagine an aeroplane, on board

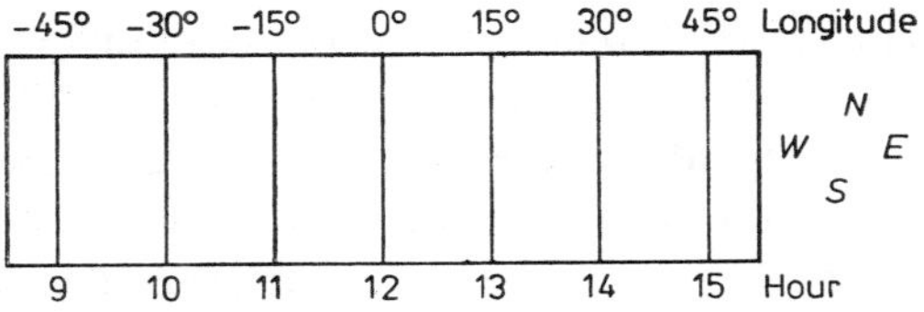

Fig. 3.6

which all watches are set for GMT. The pilot of the plane will find that his time is expressed in terms of the time on Earth and the longitude according to the formula

pilot's time (in hours) = Earth time (in hours)—1/15 of longitude (in degrees).

We can see that the relationship between the pilot's time (GMT), Earth time and longitude is linear. If the plane's flight is rectilinear and uniform from the point of view of the watches on Earth, then the pilot will note that, from the viewpoint of his watch, the flight is also uniform. Of course the reason for this conformity is that the relationship between the pilot's time and the time on Earth is linear.

So far we have looked at structures and concepts common to all theories of spacetime. Now we shall consider those properties which distinguish the different theories.

To begin with, the pre-relativistic theories assume that among many possible times (defined, for example, as the fourth coordinate of an event with respect to a given repère) there is one time especially suitable for the description of physical phenomena; it is called *absolute time.* To have this time everywhere in the Universe, we would have to synchronize all possible clocks. How can this be done? Before Einstein, little thought was given to this question, and those who did consider it, probably thought that it could be done by means of some instantly propogating signals of as-yet-unknown description. Ever since 1675, when Olaf Römer measured the velocity of light by observation of the eclipse of Jupiter's moons, it was known that light signals have finite velocity, but it seemed that something, e.g. gravitational disturbances, propagated instantly.

What is absolute time? Given a repère $(\mathfrak{o}, e)$ in our spacetime, we can call the fourth coordinate of an event the time with respect to that repère. The statement that there exists absolute time implies that we distinguish one of the times so defined.

The simultaneity of two events with respect to a repère $(\mathfrak{o}, e)$ is easily defined: events p and q $(p, q \in E)$ are simultaneous with respect to repère $(\mathfrak{o}, e)$ if $\xi^4(p) = \xi^4(q)$. A subset of E consisting of all simultaneous events is a hyperplane in E. To different values of the coordinate ξ^4 there correspond different parallel hyperplanes.

Saying that there exists absolute time amounts to saying that these hyperplanes exist objectively in spacetime, i.e. that they constitute an additional geometrical structure, regardless of whether we consider repères or not. These hyperplanes are sets of absolutely simultaneous events, or of constant absolute time t. Of course, the existence of absolute time distinguishes some of the repères, namely those whose fourth coordinate coincides with absolute time, i.e. for every event $p \in E$ we have $\xi^4(p) = t(p)$. The basis vectors e_1, e_2 and e_3 then lie in the hyperplane of the events simultaneous with $\mathfrak{o}$, as shown in Fig. 3.7. The vector e_4 has intentionally been drawn not perpendicular to this

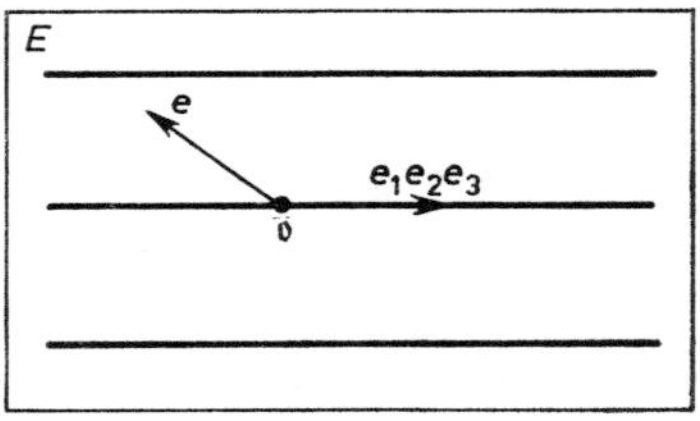

Fig. 3.7

plane because we have not introduced the concept of perpendicularity in this context. If we did so, we would have to distinguish straight lines which intersect the hyperplanes of simultaneous events at right angles. Then it could be said that the material points for which these lines are world-lines are at absolute rest.

According to the Galilean principle of relativity this statement makes no sense because all inertial systems are fully equivalent for describing physical phenomena. In contrast, absolute rest did make sense in Aristotle's theory of space and time. There, having introduced the notion of vectors perpendicular to the hyperplanes of simultaneous events, we could assign to every point in space-time a point on one of those hyperplanes (arbitrarily chosen) by moving along a straight line perpendicular to it. Spacetime would then become a Cartesian product of space and time, as pictured in Fig. 3.8.

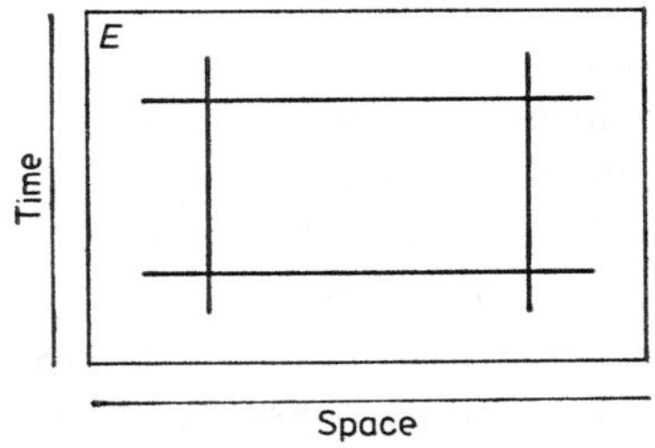

Fig. 3.8

Let us give an example illustrating the difference between the Aristotelean and Galilean theories. In either theory it is possible to find how much time elapsed between the battle of Hastings and the conception of the theory of relativity. According to Galileo, we cannot tell what is the spatial distance between these events, because we would first have to say against which reference frame we are going to measure it. An Earth-bound system is not good because it is not inertial, whilst distances measured in different inertial systems will also be different. According to Aristotle, however, there is a system among inertial systems which is "at rest", and it is with respect to that system that we should measure the distance, which will then have an absolute meaning.

Absolute time alone does not determine the structure of Galilean spacetime. In addition, it is asserted that the hyperplanes of simultaneous events are Euclidean spaces. This permits measuring distances between simultaneous events.

We shall now define the geometrical structure of Galilean spacetime in a more mathematical manner. *Galilean spacetime* is a four-tuplet (E, V, τ, h). Here (E, V) is a four-dimensional affine space and τ is a form on V, i.e. a linear mapping $\tau: V \rightarrow \mathbf{R}^1$. The space of all forms on a vector space V is usually denoted by V^*; we shall therefore write $\tau \in V^*$. Given a form τ, we can define the set $S = \{v \in V: \tau(v) = 0\}$, which is a subspace of V and the elements of which will be called the spatial vectors. Conversely, a three-dimensional subspace S of V defines—up to a constant factor—a form τ on V, which vanishes on S. Finally, h is a positive-definite product in S, i.e. a bilinear, symmetric and positive-definite mapping h: $S \times S \rightarrow \mathbf{R}^1$.

Let $\mathfrak{o} \in E$; then the absolute time t of a point p, measured from point $\mathfrak{o}$, is $t(p) = \tau(p - \mathfrak{o})$. By adding to p a spatial vector, we obtain an element in E simultaneous with p. The hyperplane of events simultaneous with p is the set $p + S = \{p + s: s \in S\}$. The scalar product h serves to measure the distance between simultaneous events. Let p and q be two such events, i.e. $(p - q) \in S$. We define the distance between p and q as the number

$$|p-q|_h = \sqrt{h(p-q, p-q)}.$$

It follows that speaking about distance only makes sense for simultaneous events, because the difference of non-simultaneous events is not a spatial vector and the scalar square $h(p-q, p-q)$ is then undefined.

We can now formulate the concept of inertial reference frames, which corresponds to distinguishing a certain class of coordinate systems in our mathematical model. Rectilinear coordinate systems in E are distinguished by the affine structure of spacetime; every such system has a corresponding repère. Inertial coordinate systems constitute a narrower class; they are associated with special repères called inertial repéres.

An *inertial repère* in E is a repère $(\mathfrak{o}, e)$ such that the basis e of the vector space V has the following properties:

(1) $\tau(e_4) = 1$,

(2) $\tau(e_\alpha) = 0$ for $\alpha = 1, 2, 3$,

(3) $h(e_\alpha, e_\beta) = \delta_{\alpha\beta}$.

$\delta_{\alpha\beta}$ is Kronecker's symbol, i.e. it is zero when $\alpha \neq \beta$ and one when $\alpha = \beta$. The vectiors e_α are spatial, mutually prependicular and of length 1. Owing to conditons (1) and (2), the coordinate ξ^4 is equal to the absolute time t measured from point $\mathfrak{o}$, i.e., for every $p \in E$ we have $\xi^4(p) = t(p)$.

A *Galilean transformation* is a one-to-one mapping of E onto itself which preserves the structure of E, that is the affine structure, τ and h. More precisely, it is a pair of one-to-one mappings $f: E \to E$, $\gamma_f: V \to V$ such that

(1) if $p \in E$ and $u \in V$ then $f(p+u) = f(p) + \gamma_f(u)$,

(2) γ_f is a linear mapping,

(3) $\tau(\gamma_f(u)) = \tau(u)$ for $u \in V$,

(4) if $v, w \in S$, then $h(v, w) = h(\gamma_f(v), \gamma_f(w))$.

We will now find the exact form of Galilean transformations. Let us decompose the point $p \in E$ with respect to the repère $(\mathfrak{o}, e)$:

$$p = \xi^\alpha e_\alpha + t e_4 + \mathfrak{o}.$$

The Galilean transformation gives

$$f(p) = \xi'^\alpha e_\alpha + t' e_4 + \mathfrak{o}.$$

We want to find the relationship between the coordinates (ξ'^α, t') and (ξ^α, t). Since e is a basis in V, we have

$$\gamma_f(e_\alpha) = R^\beta_\alpha e_\beta + Q_\alpha e_4,$$

$$\gamma_f(e_4) = V^\alpha e_\alpha + P e_4.$$

From condition (3) it follows that firstly $Q_\alpha = 0$ and secondly $P = 1$. Property (4) implies orthogonality of the matrix R^α_β, i.e.

$$R^\alpha_\beta R^\gamma_\delta \delta_{\alpha\gamma} = \delta_{\beta\delta}.$$

Next we use condition (2), the linearity of γ_f:

$$f(p) = f(\xi^\alpha e_\alpha + te_4 + \mathfrak{o}) = \gamma_f(\xi^\alpha e_\alpha + te_4) + f(\mathfrak{o})$$
$$= \xi^\alpha R^\beta_\alpha e_\beta + tV^\beta e_\beta + te_4 + \big(f(\mathfrak{o}) - \mathfrak{o}\big) + \mathfrak{o}\,.$$

Since $f(\mathfrak{o}) - \mathfrak{o}$ is a constant vector,

$$f(\mathfrak{o}) - \mathfrak{o} = a^\alpha e_\alpha + t_0 e_4\,,$$

we obtain

$$f(p) = (\xi^\alpha R^\beta_\alpha + tV^\beta + a^\beta) e_\beta + (t + t_0) e_1 + \mathfrak{o}\,.$$

Thus, the Galilean transformations have the form

$$\xi'^\alpha = R^\alpha_\beta \xi^\beta + V^\alpha t + a^\alpha,$$

$$t' = t + t_0\,.$$

Having developed the geometrical model of spacetime, we can formulate the laws of mechanics of material points, rigid bodies and continua in its language. For example, the equation of free motion will have the form

$$\frac{d^2 x^i}{dt^2} = 0,$$

where t is absolute time. It then turns out that Galilean transformations carry solutions of the equations of mechanics into solutions of the same equations; in other words, Galilean transformations are symmetries in mechanics. The Galilean relativity principle is often expressed by saying that all inertial systems are fully equivalent for the description of mechanical phenomena. Of course this does not mean that every physical phenomenon manifests itself identically in all inertial systems. For example, the phenomenon of a material point resting in a certain inertial reference frame will be seen as a rectilinear and uniform motion in some other inertial frame. This agrees with the Galilean relativity principle, because both rectilinear uniform motions and rest with respect to an inertial frame are solutions of the free motion equation mentioned above.

Historically, the first fully geometrical formulation of mechanics was given by Cartan [54]. The problem was later studied by many other authors.

CHAPTER 4

In Search of the Ether

So far in our discussion of Galilean spacetime we were only concerned with mechanical phenomena. The question arises: how should we deal with electromagnetic phenomena, particularly optical phenomena? We have singled out the optical phenomena because, historically, their role was especially important.

Consider the propagation of light in a vacuum. It is known to be rectilinear, like the free motion of material points. There is one significant difference however: the velocity of light is constant. More precisely, in mechanics the initial conditions, i.e. the position $\mathbf{r}(0)$ and velocity $\dot{\mathbf{r}}(0)$, determine the position $\mathbf{r}(t)$ of the point at any subsequent time t. In optics, on the other hand, it is enough to know $\mathbf{r}(0)$ and the direction of the initial velocity to know the position $\mathbf{r}(t)$ of the light point at time t.

To formulate the laws of optics in conformity with the Galilean model, it is first necessary to distinguish the reference frames with respect to which the velocity of light is c. As we know, the velocity of light appears in Maxwell's equations governing electromagnetic phenomena. Let us consider what the velocity is with respect to an inertial system.

If the curve $t \mapsto x(t) \in E$ is the world-line of a material point or a light ray, then $\mathrm{d}x/\mathrm{d}t$ is called the *four-velocity* of that point (or ray). In every inertial reference frame the coordinates of the velocity four-vector are $(\mathrm{d}x^i/\mathrm{d}t) = (\mathrm{d}\mathbf{r}/\mathrm{d}t, 1)$. Hence $\mathrm{d}x/\mathrm{d}t - e_4$ is a spatial vector (velocity in the ordinary sense) and we can measure its length $|\mathrm{d}x/\mathrm{d}t - e_4|_h$ as defined through the scalar product h in the space S. Let u be the four-velocity of a light ray. All inertial frames for which the equality $|u - e_4|_h = c$ holds for every light ray will be distinguished. Since light can propagate in all directions, it is easy to show that there exists only one such vector e_4. We shall call it the *ether* and denote it by e.

Let us stress the double meaning of the word ether. Firstly, vector e determines the inertial frame (called the ether frame), relative to which the velocity of light is c, and in which, consequently, Maxwell's equations are satisfied. Secondly, the term ether has been used to mean a hypothetical material medium at rest in that frame; electromagnetic phenomena were understood to be manifestations of vibrations of that medium. Notice that the existence of a material ether implies the existence of the geometrical ether, but not vice versa.

It is possible to conceive of a physical theory which would deny the existence of the material ether and at the same time admit an inertial frame with respect to which the velocity of light is c.

With a greater emphasis placed on the material aspect of the existence of ether, it is easy to come to a certain generalization. We can imagine an ether which is not at rest in any inertial frame. Its velocity changes from point to point in spacetime. From a mathematical viewpoint we then deal with a vector field $E \ni p \mapsto e(p) \in V$, the four-velocity field of the material ether. In this way, at every point of spacetime we have a local inertial frame whose fourth vector coincides with e (Fig. 4.1).

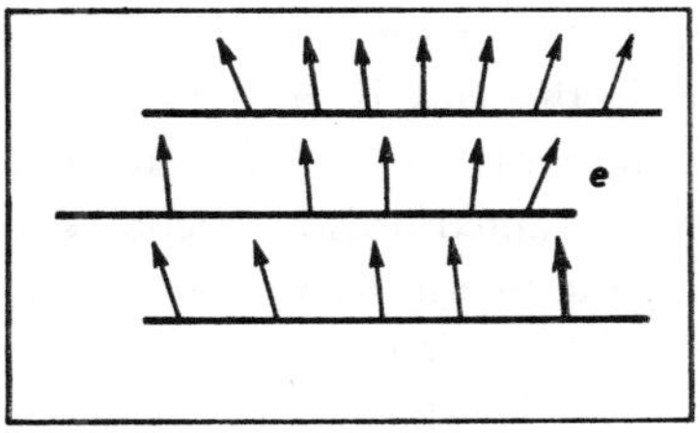

Fig. 4.1

The theories in which the vector e is variable are referred to as theories of convected, or dragged, ether. The term comes from the once plausible assumption that moving ponderable bodies drag along some or even the whole of the ether that surrounds them. Theories in which e is a constant vector are called the theories of absolute ether.

We shall now look back on some of the experimental attempts to detect ether and reveal its nature.

At first sight it appears that a strong argument in favour of the existence of an absolute ether is provided by the phenomenon of aberration. Let us recall its principle. Light from a star G on the axis of rotation of the Earth

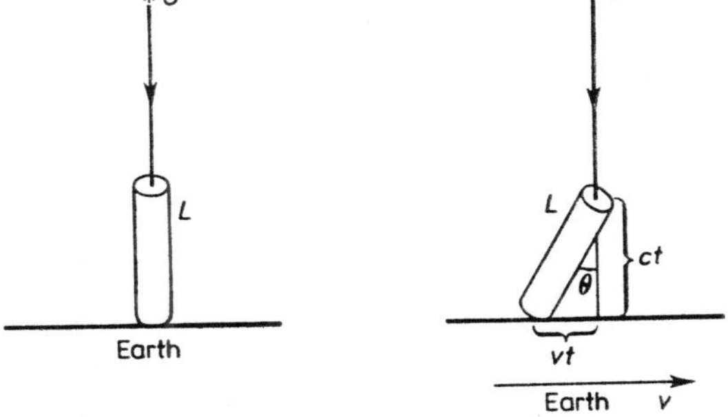

Fig. 4.2

about the Sun is obverved by means of a telescope L. The picture on the left of Fig. 4.2 illustrates the case where the Earth is at rest with respect to the hypothetical ether. On the right, the Earth moves with velocity v relative to the ether. To observe the light emitted by the star G, the observer must tilt the telescope by an angle θ such that $\tan\theta = v/c$). Observations yielded the angle 2θ by which the position of the telescope had to be altered every half a year if the star was still to be seen in it. The velocity v obtained from the above formula turned out to be approximately 30 km/s, which is the orbital velocity of the Earth relative to the Sun.

If we accept the existence of an absolute ether and the corpuscular nature of light, the phenomenon of aberration becomes analogous to that of slanting streams formed by raindrops on the windows of a fast moving train. The counter-part of the ether in this case is the air which the train moves past.

This analogy, although attractive, is not complete, and as it is, the effect of aberration can also be explained without recourse to the concept of ether.

Fizeau (1859) experimented with the velocity of light in moving media. He transmitted light through tubes filled fast moving water, both in the direction of the flow and opposite to it. He then compared the velocity of light in each case. It turned out that the two velocities differed, which suggested that the ether was, at least partially, dragged by the water. Quantitatively, good results are obtained if we assume, after Fresnel, that the coefficient of ether drag for a material medium with a refractive index n is $\alpha = 1-1/n^2$. By the ether drag coefficient we mean the ratio of the velocity of the dragged ether to the velocity of the body that drags it. Hence for optically inert bodies we have $\alpha = 0$, which agrees with the phenomenon of aberration, while for bodies optically active $\alpha > 0$, i.e. such bodies will partially drag the ether.

A more detailed analysis shows that the optical phenomena which can be described by formulae involving quantities of order not higher than first in $\beta = v/c$ are easily tractable by many different theories. In particular, the theory of relativity gives the same predictions for effects of this type, including the change in the velocity of light in moving water (Fizeau's experiment) and

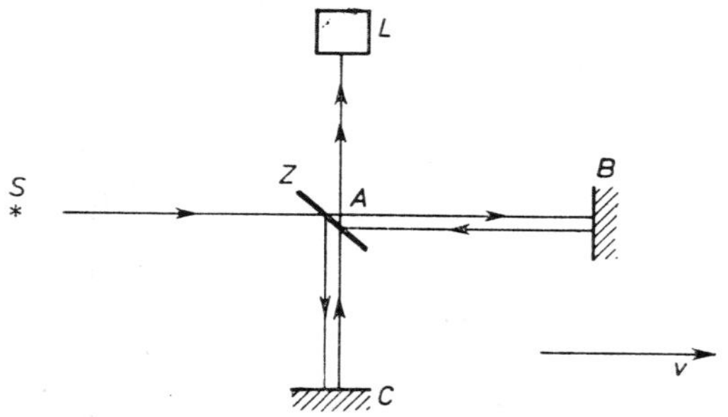

Fig. 4.3

aberration, among others. In this connection, effects of the second order were investigated as possible arguments in favour of the ether theory.

A relevant experiment (first suggested by Maxwell) was carried out in 1887 by Michelson and Morley [34]. The beam of light from source S falls on a semi-transparent plate Z and is partially reflected and partially transmitted, to reach the mirrors B and C. From there, the light is reflected back to L, where its interference pattern is observed (Fig. 4.3). The apparatus is positioned in such a way that the arm AB is parallel to the direction of the velocity of the Earth relative to the Sun (and therefore, approximately, relative to the ether, if the Sun-bound frame approximates the inertial frame of the ether). If l_1 is the distance between the point A and the mirror B, then the time taken by the light to travel along the path ABA is

$$\frac{l_1}{c-v}+\frac{l_1}{c+v}=\frac{2l_1}{c}\,\frac{1}{1-\beta^2}.$$

From the triangle ACA (Fig. 4.4), which represents the path of the ray in

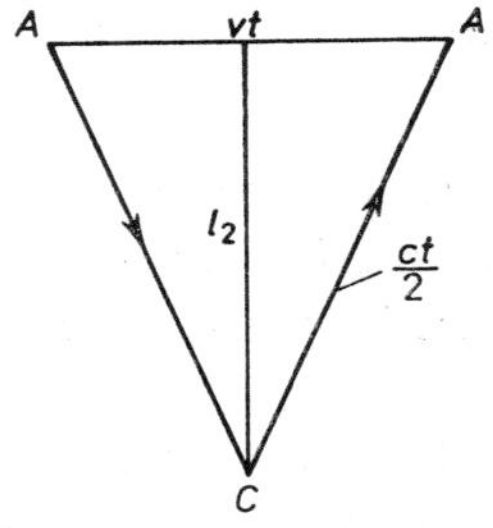

Fig. 4.4

the supposed ether frame, we find that the time t along the path ACA is

$$\frac{2l_2}{c\sqrt{1-\beta^2}}.$$

The difference between the two times then is

$$\Delta=\frac{2}{c\sqrt{1-\beta^2}}\left(\frac{l_1}{\sqrt{1-\beta^2}}-l_2\right).$$

If the interferometer is rotated so as to interchange the roles of the mirrors B and C, the time difference is calculated as

$$\Delta'=\frac{2}{c\sqrt{1-\beta^2}}\left(l_1-\frac{l_2}{\sqrt{1-\beta^2}}\right).$$

The observations of Michelson and Morley aimed at finding whether this rotation produces a shift in the interference pattern, i.e. at experimentally measuring the difference

$$\Delta-\Delta' = \frac{2(l_1+l_2)}{c\sqrt{1-\beta^2}}\left(\frac{1}{\sqrt{1-\beta^2}}-1\right) \cong \frac{l_1+l_2}{c}\beta^2.$$

To a high degree of accuracy and independent of the time of year, the result was $\Delta-\Delta' = 0$. The measurements have since been repeated many times. The result of highest experimental accuracy was obtained in 1958 by Townes and his collaborators [6] with the use of ammonia masers. Their measurements gave $\beta < 10^{-7}$, while the theory of the absolute ether predicts a value of β of the order of $\dfrac{30 \text{ km/s}}{300000 \text{ km/s}} = 10^{-4}$.

In 1892, to explain the null result of the Michelson–Morley experiment, Fitzgerald and Lorentz independently suggested that bodies moving relative to the ether undergo a contraction in their direction of motion by a factor of $\sqrt{1-\beta^2}$. Indeed, if we accept the Fitzgerald–Lorentz hypothesis, then

$$\Delta = \Delta' = \frac{2}{c\sqrt{1-\beta^2}}(l_{10}-l_{20}),$$

where l_{10} and l_{20} denote the lengths of the arms of the interferometer at rest with respect to the ether. The interferometer used by Michelson and Morley had arms of nearly equal length, $l_{10} \simeq l_{20}$, which explains the very small absolute value of the shift Δ.

A modification of the Michelson–Morley experiment, aiming at a verification of the Fitzgerald–Lorentz hypothesis, was designed by Kennedy and Thorndike [30]. They used an interferometer with arms of different lengths, $l = l_{10} \gg l_{20}$. In this case, assuming the contraction hypothesis,

$$\Delta(\beta) \approx \frac{2l}{c}\left(1+\frac{1}{2}\beta^2\right).$$

Kennedy and Thorndike compared the values of $\Delta(\beta)$ obtained in the same experimental setting at different times of day and year. If the ether existed and the contraction hypothesis were true, then, owing to the motion of the Earth, the value of $\Delta(\beta)$ should depend on the time of measurement, but this effect was not observed. Thus, the Lorentz–Fitzgerald hypothesis is not sufficient to explain the experiment of Kennedy and Thorndike.

An attempt to save Newtonian physics which was comparatively difficult to refute was Ritz's emission theory [42], published in 1908, when the theory of relativity had already been formulated. It was a modernized version of the corpuscular theory, stating that the velocity of light in vacuum is c relative to the source. In other words, every source has its own ether. The theory explained both the effect of aberration and the null results of the Michelson–Morley experiments.

The followers of Ritz's theory gave different answers to the question of what happens to the light reflected from a mirror that moves relative to the source. Ritz himself maintained that the velocity of light is still c with respect to the original source.

The emission theory could not be reconciled with modern quantum views on the mechanism of the emission of light. In addition, it is contradicted by the observations of the Doppler effect on binary stars.

The question arises: have all possible ways of reconcilling electrodynamics with the Galilean model already been exhausted? Aren't there any models other than the theory of relativity in which one could develop electrodynamics? It is not easy to give a categoric answer to these questions.

To illustrate what we mean, consider the following situation in data analysis. Suppose that a physicist obtains a sequence of numbers which he plots on a graph. Someone who knows what's what will notice that the points delineate an exponential curve and, using the methods of data analysis, will find that an exponential curve is indeed a good description of the data. Someone else, however, will say that he only accepts algebraic curves and that he does not like the analysis produced by his predecessor. After all, if we have a finite number of experimental points, we can always choose a polynomial of a suitable order which will fit the data (Fig. 4.5).

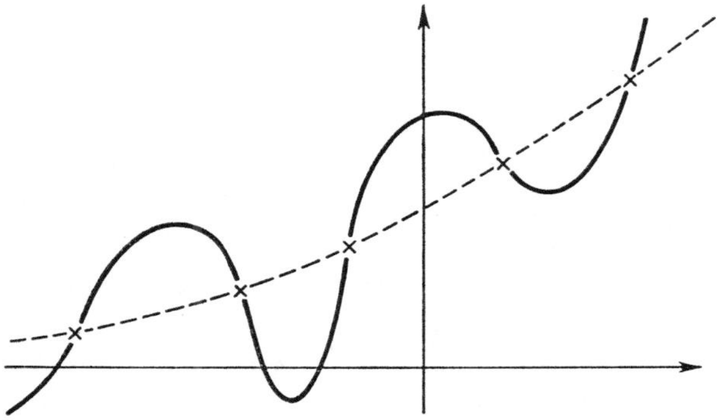

Fig. 4.5

There is a loose analogy between this situation and the relationship between theory and experiment. A finite number of experimental facts can always be explained by many theories. One of the criteria for accepting one of them is simplicity. Another, much more important criterion is whether it can be used for modelling other experiments. And indeed, the distinguishing feature of the theory of relativity is that not only does it explain the Michelson–Morley, the

Kennedy–Thorndike, the aberration and other classical experiments of optics, but it is also a basis for other physical theories, which describe a much richer class of physical phenomena. Thus, if someone puts forward a theory explaining the optical experiments of the turn of the 19th century, he should next verify that his theory is in agreement with the results of quantum electrodynamics and elementary particle physics.

At the time he published his paper *On the electrodynamics of moving bodies*, Einstein [7] did not know the results of the Michelson–Morley experiments. His considerations were based on the approximate relativity principle for electromagnetic phenomena, which had been known for a long time. Everybody knows that by moving a magnet near a closed conductor we will generate an electromotive force, and hence a flow of current in the conductor. The direction and magnitude of the current will not change if instead of moving the magnet we move the conductor in the opposite direction with the same speed (Fig. 4.6). These two situations can be interpreted as the same phenomenon

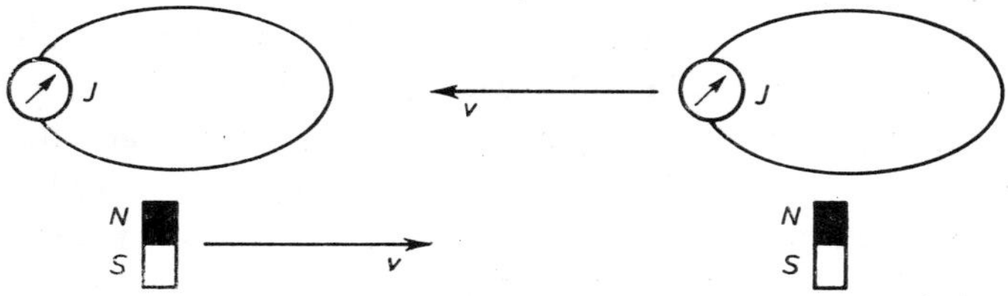

Fig. 4.6

observed in two different inertial frames. The effect observed does not depend on the choice of the inertial frame.

Einstein, who heard about the failure of the experiments trying to detect motion relative to the ether, found a way out of the dillema: his *principle of relativity*. It says: *mechanical and electromagnetic phenomena are the same in all inertial frames*. This principle generalizes the Galilean relativity principle in that it embraces not only mechanical but also electromagnetic phenomena. Nowadays, the principle of relativity is formulated in a slightly different way to say that every physical phenomenon is the same in all inertial frames. At the turn of the 19th century effectively only mechanical and electromagnetic phenomena were known, so that the two formulations were then equivalent.

It may appear that the step made by Einstein was not so big, that it was a formalization of the facts known for a long time. As it turned out, however, it had far-reaching consequences. For example, the relativity principle implies that Maxwell's equations should have the same form in all inertial frames and consequently the velocity of light should have the same value c in all inertial systems. This of course contradicts the Newtonian theorem of the addition of velocities.

The principle of relativity is often expressed in terms of inertial observers, who are a kind of anthropomorphization of inertial frames. It is indeed convenient to use, besides a repère ($\mathfrak{o}$, e_1, e_2, e_3, e_4), an observer whose world-line is tangent to the vector e_4. The observer is equipped with a clock which measures a unit of time while he passes from the point $\mathfrak{o}$ to the point $\mathfrak{o}+e_4$. In addition, he has theodolites "levelled" along the vectores e_1, e_2, and e_3, and can measure the angle of incidence of the light rays that reach him.

Einstein's relativity principle says that physical phenomena are the same for all inertial observers. In the Galilean approach, all inertial observers had equal rights with respect to mechanical phenomena. In Einstein's, we have the principle of their full equivalence with respect to all phenomena (Fig. 4.7). This

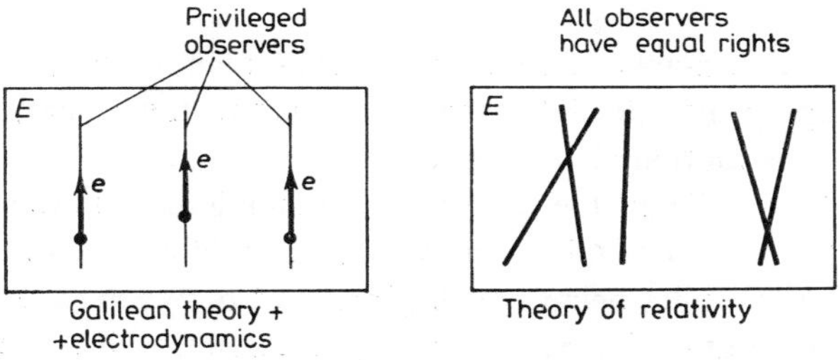

Fig. 4.7

principle, together with a few additional assumptions of a mathematical nature, has led to significant changes in the ways of describing the world.

CHAPTER 5

Predictions of the Theory of Relativity and Their Experimental Verification

Accepting the principle of relativity implies relinquishing absolute time. For, let us suppose that absolute time exists and that the velocity of light does not depend on the motion of the source.

Let a point $p \in E$ lie on the hyperplane of simultaneous events defined by the equation $t = t_0$. From this point we send two light beams in opposite directions, their world-lines being rays beginning at p. We pass a straight line through the midpoint of the segment joining the points of intersection of these rays with a hyperplane $t = t_1$, where $t_1 > t_0$, and through p (Fig. 5.1). This

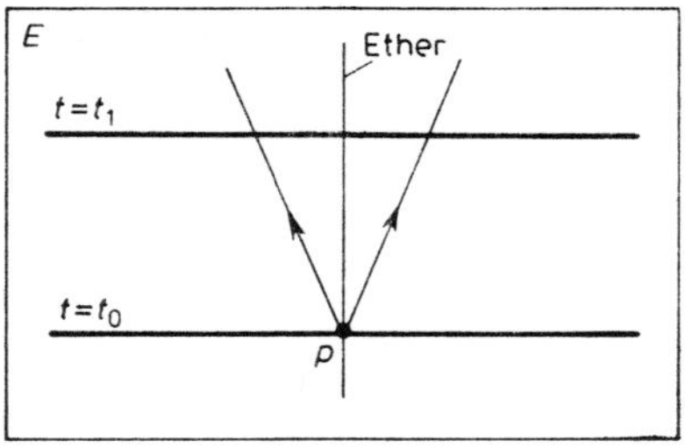

Fig. 5.1

straight line is chosen canonically, since it is independent of the motion of the source (owing to the isotropy of the space, it does not depend on the direction of the beams either). The material point for which this line is the world-line is naturally described as being at rest. Thus we come to the notion of absolute rest (ether), a contradiction of the principle of relativity.

From now on, therefore, we shall no longer assume that spacetime E incorporates absolute time. We shall maintain the affine space implied by Newton's 1st law.

Every inertial observer has a clock. The time of an event p relative to an observer O associated with the repère $(\mathfrak{o}, e_\alpha, e_4)$ (where $\alpha = 1, 2, 3$) is defined by the equality

$$p = t(p)e_4 + \xi^\alpha(p)e_\alpha + \mathfrak{o},$$

while the time of the same event p with respect to an observer O' is defined by

$$p = t'(p)e'_4 + \xi'^\alpha(p)e'_\alpha + \mathfrak{o}'.$$

The times t and t' are connected with the bases and do not have any metric properties (Fig. 5.2). The question arises: how can the clocks of different

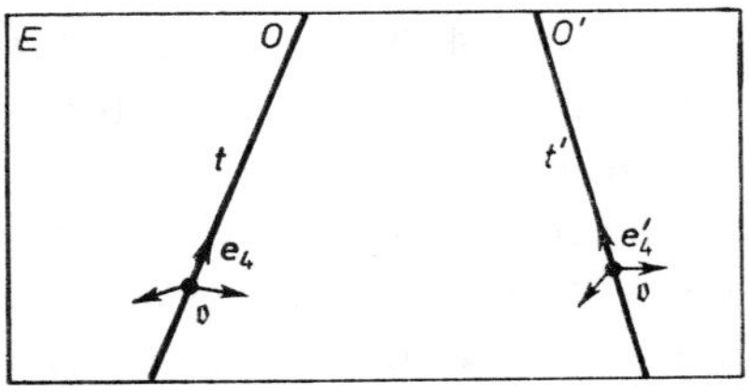

Fig. 5.2

observers be synchronized or, in other words, what does it mean in this model to say that observer O and O' use identical clocks?

First consider the case where the world-lines of the observers intersect (Fig. 5.3). To simplify the argument, assume that the repère of either observer

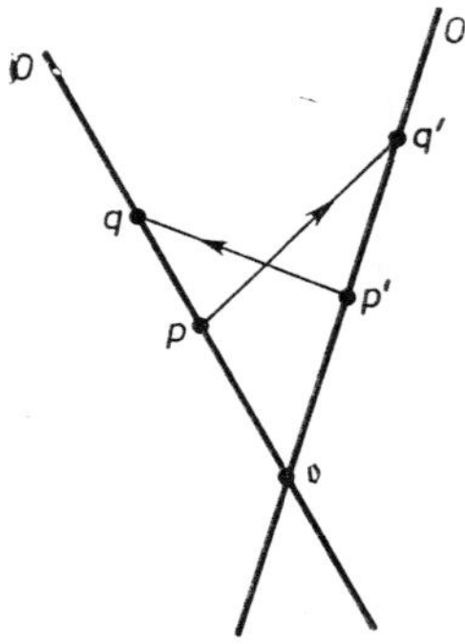

Fig. 5.3

is attached at the point $\mathfrak{o}$ of intersection of the world-lines. Then $t(\mathfrak{o}) = t'(\mathfrak{o}) = 0$ The segments pq' and $p'q$ represent the light rays. If $t(p) = t'(p')$, then since the observers have equal rights and their clocks are identical, we should have $t(q) = t'(q')$. Let us stress that this conclusion would not hold if we adopted the ether hypothesis, letting the velocity of light depend on the reference frame.

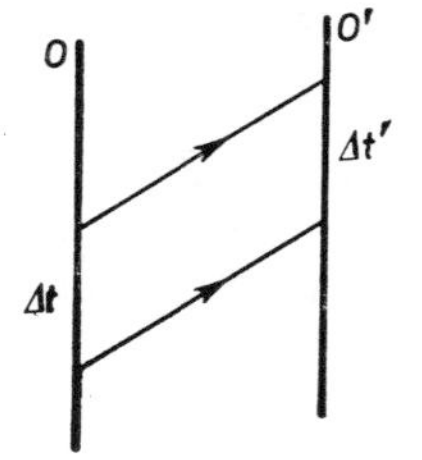

Fig. 5.4

Figure 5.4 illustrates the synchronization problem when the world-lines of observers O and O' are parallel. The agreement between the clocks now means that $\Delta t = \Delta t'$. Of course in this case there is no good way of defining a common starting point from which to measure time.

If the world-lines of observers O and O' are skew, the clocks can be synchronized in two steps—there exists a world-line (of an observer O'') intersecting one of the given lines and parallel to the other. From now on we shall assume that the clocks of all inertial observers are synchronized as described above.

The relativity of simultaneity. Since we rejected the concept of absolute time, we must also give up absolute simultaneity. Two events regarded as simultaneous by one observer will not, in general, be simultaneous to another. We shall analyse this question in a simple case.

Let the world-lines of observers O and O' intersect. Observer O sends intermittent light signals towards O' (Fig. 5.5). The triangles with vertices

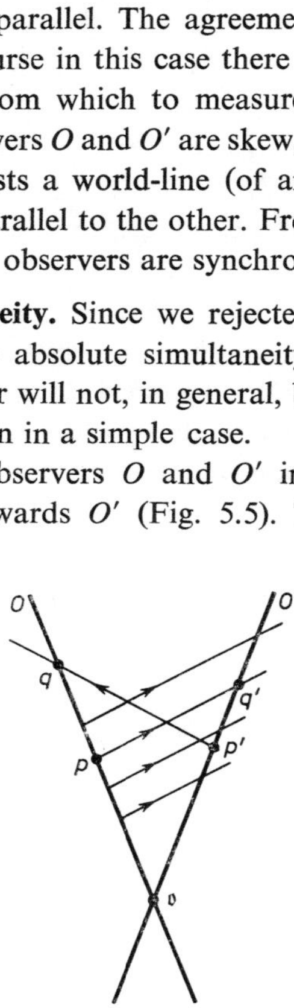

Fig. 5.5

at the points of emission by O, reception by O', and the intersection $\mathfrak{o}$, are similar. It follows that the coefficient α in the equation $t'(q') = \alpha t(p)$ does not depend on p—the point from which the signal is sent. Similarly, we have $t(q) = \alpha' t'(p')$. By chossing the events p and p' so that $t(p) = t'(p')$ and using the synchronization condition, we conclude that $\alpha' = \alpha$. Obviously, $\alpha \geqslant 1$.

Now suppose that a signal is sent by observer O at a time t_1 and is received by O' at a point p' corresponding to time t'. Using a mirror, observer O' reflects the signal back to O, who records its arrival at an instant t_2 (Fig. 5.6)

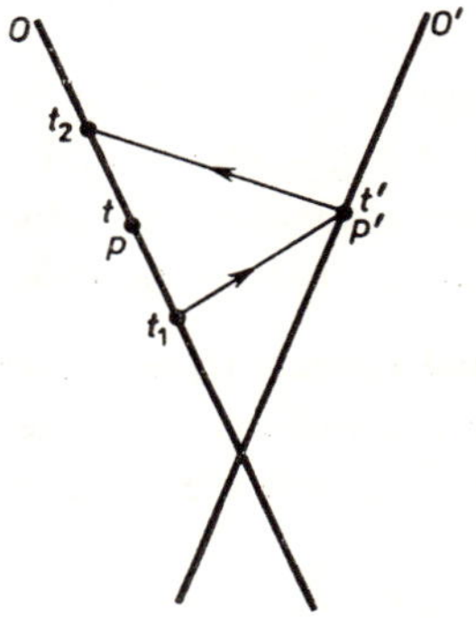

Fig. 5.6

Which event on O's world-line will be regarded by O as simultaneous with the event p'? From the point of view of observer O, the sensible choice is the event p such that

$$t = t(p) = \frac{1}{2}(t_1 + t_2) = \frac{1}{2}\left(\alpha + \frac{1}{\alpha}\right)t'.$$

It is easy to see that $t > t'$. This effect is known as time dilation. Observer O will comment on this fact by saying that the clock of O' is slow. Observer O' will not agree to regard the events p and p' as simultaneous. As an exercise we propose that the reader finds the event on O's world-line which will be viewed by O' a simultaneous with p', and shows that to O' the clock of observer O will appear to have slowed down by the same factor. We conclude that moving clocks go at a slower rate than clocks at rest and that every observer can notice this phenomenon.

Let us calculate the coefficient α. If O' moves with a velocity V relative to O, the distance between O' and O at the time $t = \frac{1}{2}(t_1 + t_2)$ is

$$Vt = \frac{1}{2}V(t_1 + t_2) = \frac{1}{2}(1 + \alpha^2)Vt_1.$$

To travel the same distance, light will need the time $\frac{1}{2}(t_2 - t_1) = \frac{1}{2}(\alpha^2 - 1)t_1$, so

$$\frac{1}{2}(1 + \alpha^2)Vt_1 = \frac{1}{2}(\alpha^2 - 1)ct_1.$$

After simple transformations we obtain

$$\alpha = \sqrt{\frac{1+\beta}{1-\beta}}, \qquad \text{where } \beta = \frac{V}{c}.$$

Hence the time dilation is given by

$$t' = \sqrt{1-\beta^2}t.$$

The same formula, with t and t' replaced by Δt and $\Delta t'$, holds for inertial observers whose world-lines do not intersect. In the special case where the world-lines of O and O' are parallel, i.e. the observers are at rest relative to each other, we have $\Delta t' = \Delta t$, i.e. there is no time dilation.

Distance measurements and Lorentz transformation. Suppose that observers O and O' want to determine the position and time of occurrence of an event p. For simplicity, let us assume that p is in the same plane as the world-lines of the observers.

Figure 5.7 represents the radar method of measuring distances. At a time t_1

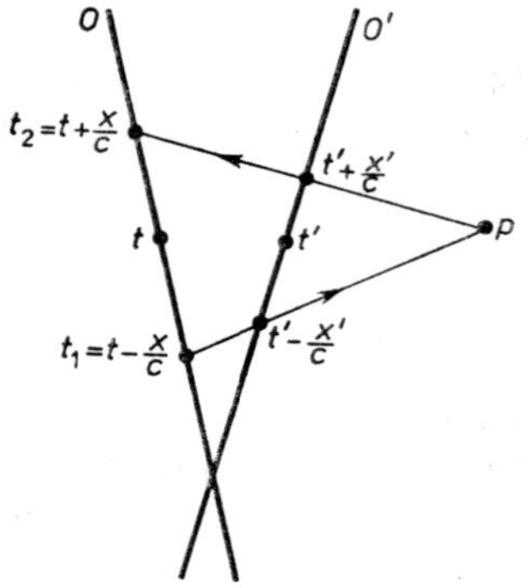

Fig. 5.7

observer O sends a light signal which passes O', is reflected at p, again passes O', and returns to O at time t_2. Observer O will calculate the distance of the event p as

$$x = \frac{1}{2}c(t_2-t_1)$$

and the instant when it occurs as

$$t = \frac{1}{2}(t_1+t_2).$$

An analogous expression will be obtained by observer O'. As a result, the times at which the light meets the observers will have the values shown in Fig. 5.7. According to our previous considerations we have

$$t'-\frac{x'}{c} = \alpha\left(t-\frac{x}{c}\right),$$

$$t+\frac{x}{c} = \alpha\left(t'+\frac{x'}{c}\right).$$

It follows that $c^2t'^2-x^2 = c^2t^2-x^2$. Thus the expression $c^2t^2-x^2$ is independent of the reference frame. We say that it is an invariant in special relativity theory.

In terms of t and x, t' and x' are expressed as

$$t' = \frac{1}{2}\left(\alpha+\frac{1}{\alpha}\right)t-\frac{1}{2}\left(\alpha-\frac{1}{\alpha}\right)\frac{x}{c},$$

$$x' = \frac{1}{2}\left(\alpha+\frac{1}{\alpha}\right)x-\frac{1}{2}\left(\alpha-\frac{1}{\alpha}\right)ct.$$

Substituting the expression we found for α, involving the velocity V of observer O' relative to observer O, we obtain

$$t' = \frac{t-(Vx/c^2)}{\sqrt{1-\beta^2}}, \qquad x' = \frac{x-Vt}{\sqrt{1-\beta^2}}.$$

The above formulae are called the *special Lorentz transformation equations*. They were first found by Larmor [32] as transformations which do not change the form of Maxwell's equations; his approach, however, was purely formal. It was Einstein who first gave a physical meaning to the quantities involved, i.e. interpeted them as corresponding to the readings of good clocks and correct measurements of distances.

We shall now discuss some consequences of the special Lorentz transformation.

Non-relativistic limit. Suppose that the velocity V of the observer O' is small compared to the velocity of light c, that is $\beta \ll 1$. The Lorentz transformation equations then reduce to

$$t' = t, \qquad x' = x-Vt.$$

The same result is obtained by taking the limit as $c \to \infty$. The resulting formulae correspond to the special Galilean transformation. For speeds small compared with c, or, equivalently, if c is considered infinitely large, the results of relativistic physics become the results of Newtonian physics. This statement refers not only to the Lorentz transformation but to all results obtained on the basis of the theory of relativity.

Doppler effect. Consider a plane monochromatic wave moving from observer O to observer O'. The field component of such a wave is proportional to $\cos k(x-ct)$. The frequency of the wave multiplied by 2π is $\omega = kc$, and the wavelength is $\lambda = 2\pi/k$. We know that $x-ct = \frac{1}{\alpha}(x'-ct')$, so that $\omega' = \frac{k}{\alpha}c = \frac{\omega}{\alpha}$, while $\lambda' = \alpha\lambda$. Let us write the formula for the transformation of frequency as follows:

$$\omega' = \frac{1-\beta}{\sqrt{1-\beta^2}}\,\omega.$$

The numerator coincides with the classical formula while the denominator is a consequence of time dilation. The relativistic Doppler effect was observed experimentally in 1937 by Ives and Stilwell, who thus confirmed time dilation.

Length contraction. Let a rod of length l_0 in the system in which it is at rest move with a velocity V relative to an inertial observer O. To be able to say what the rod's length is as measured by observer O, we must first specify the method of measurement. Suppose that O uses the radar method, i.e. the same method which led us to the Lorentz transformation. The instants t_1 and t_2 correspond to the rod's endpoints passing O. The observer will send two light signals towards the ends of the rod, and will do this in such a way that—from his point of view—the reflection from both ends will occur simultaneously (Fig. 5.8). Using the method described previously, he will determine the distances

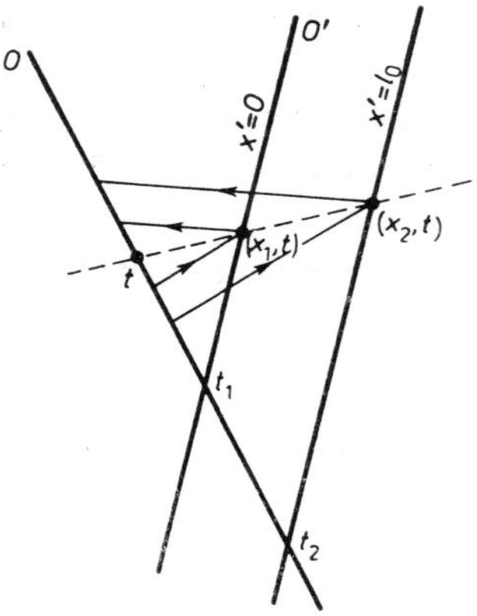

Fig. 5.8

to each end, x_1 and x_2, and will conclude that the length of the rod is $l = x_2 - x_1$. By placing another fictitious observer O' at one of the rod's endpoints and using the Lorentz transformation, we find that

$$l = l_0\sqrt{1-\beta^2}$$

Thus from the point of view of observer O the rod is shorter than it is from the point of view of O', relative to whom the rod is at rest. Let us note that the above formula for length contraction coincides with that proposed by Fitzgerald and Lorentz to explain the null result of the Michelson–Morley experiment.

Another way in which O can measure the length of the rod is to measure the velocity of the rod, V, record the times t_1 and t_2 at which the rod's ends pass him, and hence find the length of the rod as $l = V(t_1 - t_2)$. It is easy to verify that the result will be the same as in the radar method.

Velocity addition theorem. Observers O and O' describe the motion of a material point differently (Fig. 5.9). Observer O reports the distance x at time t,

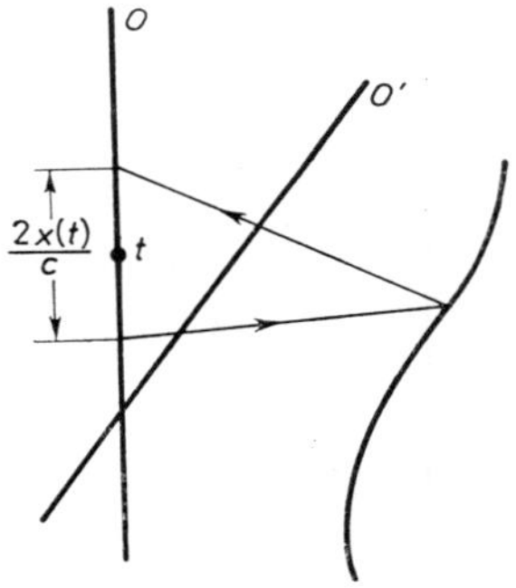

Fig. 5.9

i.e. gives a function $t \mapsto x(t)$, while observer O' uses a function $t' \mapsto x'(t')$. The velocity of the material point relative to O is $v = \mathrm{d}x/\mathrm{d}t$ whilst that relative to O' is $v' = \mathrm{d}x'/\mathrm{d}t'$. Since

$$x = \frac{x' + Vt'}{\sqrt{1-\beta^2}} \quad \text{and} \quad t = \frac{t' + \dfrac{Vx'}{c^2}}{\sqrt{1-\beta^2}},$$

we obtain

$$v = \frac{\mathrm{d}x}{\mathrm{d}t} = \frac{\mathrm{d}x' + V\mathrm{d}t'}{\mathrm{d}t' + \dfrac{V\mathrm{d}x'}{c^2}} = \frac{v' + V}{1 + \dfrac{Vv'}{c^2}}.$$

Let us look at this formula more closely. The first conclusion is that the inequality $|v'| \leqslant c$ holds if and only if $|v| \leqslant c$. Furthermore, $v = c$ if and only if $v' = c$. Besides, this was an initial assumption. The formula by no means implies that the velocity v cannot exceed the velocity of light. The theory of relativity does not make a categorical statement on this subject. The question of super-light speeds will be taken up further in Chapter 6.

Another thing to note is that in the non-relativistic limit, as $c \to \infty$, we obtain $v = v' + V$, the classical, Galilean formula for the addition of velocities

The "paradox" of twins (the clock paradox). The formula for time dilation, $t' = t\sqrt{1-\beta^2}$, leads to conclusions which are seemingly paradoxical. It follows from it, for example, that if one of twin brothers is sent for a long journey in space, he will return to Earth younger than his brother. This statement is argued against on the grounds that the situation of the two brothers is identical since each of them moves with the same (in magnitude) velocity relative to the other. The fallacy of such a reasoning is illustrated by Fig. 5.10. The fact

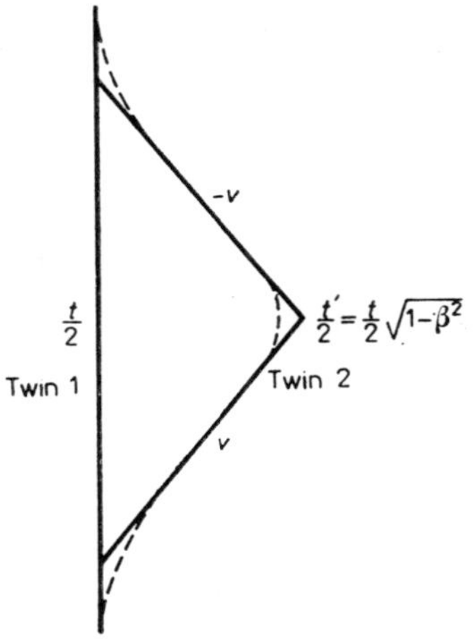

Fig. 5.10

that one of the brothers is subjected to accelerations breaks the symmetry between them. The continuous line in Fig. 5.10 is an idealization of the twin's journey; the world-line of the traveller must be smooth because hee cannot be subjected to infinitely large accelerations. His real world-line, in the periods when the engines of his rocket run, is represented by the dashed line. We shall return to this question in the next chapter.

Geometrization of velocity addition. The invariant of the special Lorentz transformation, $x^2 - c^2t^2$, resembles the Pythagorean square of a distance. Indeed, putting $\tau = ict$, we can write it as $x^2 + \tau^2$. Let ψ be the number defined by the equation $\tanh\psi = \beta$. Then

$$\cosh\psi = \frac{1}{\sqrt{1-\beta^2}}, \qquad \sinh\psi = \frac{\beta}{\sqrt{1-\beta^2}}.$$

The Lorentz transformation equations take the form

$$x' = x\cosh\psi - ct\sinh\psi,$$
$$ct' = -x\sinh\psi + ct\cosh\psi$$

Substituting $\psi = \mathrm{i}\varphi$, we obtain ($\cosh\mathrm{i}\varphi = \cos\varphi$, $\sinh\mathrm{i}\varphi = \mathrm{i}\sin\varphi$):

$$x' = x\cos\varphi - \tau\sin\varphi,$$

$$\tau' = x\sin\varphi + \tau\cos\varphi.$$

Thus the passing from one inertial frame to another in the plane (x, τ) corresponds to a rotation through angle φ. The angle ψ is called the *hyperbolic angle*. It is easy to verify that $\alpha = \sqrt{(1+\beta)/(1-\beta)} = \exp\psi$, i.e. $\psi = \ln\alpha$. Imagine three inertial observers O, O' and O'', whose world-lines intersect at one point (Fig. 5.11). The coefficients α, α' and α'' correspond to the transi-

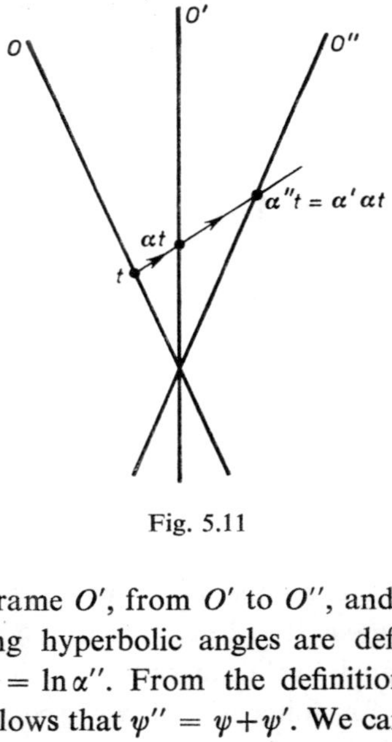

Fig. 5.11

tions from frame O to frame O', from O' to O'', and from O to O'', respectively. The corresponding hyperbolic angles are defined by the equalities: $\psi = \ln\alpha$, $\psi' = \ln\alpha'$, $\psi'' = \ln\alpha''$. From the definition of the coefficients α we have $\alpha'' = \alpha'\alpha$. It follows that $\psi'' = \psi + \psi'$. We can see that the hyperbolic angle is an additive quantity. In special relativity, instead of adding velocities when passing from one inertial frame to another we can add the hyperbolic angles. However, such a simple (commutative) addition law is only valid for motions in the same direction.

CHAPTER 6

Minkowski Geometry

We have seen that two inertial observers choosing the event $\mathfrak{o}$ corresponding to the intersection of their world-lines as the origin of the system (Fig. 6.1)

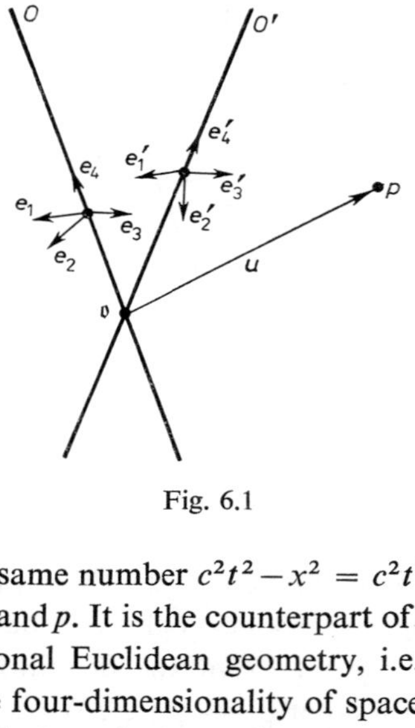

Fig. 6.1

ascribe to an event p the same number $c^2t^2 - x^2 = c^2t'^2 - x'^2$, called the square of the interval between $\mathfrak{o}$ and p. It is the counterpart of the square of the distance in ordinary two-dimensional Euclidean geometry, i.e. of an expression of the form $x^2 + \tau^2$. Taking the four-dimensionality of spacetime into account is not difficult. By a suitable choice of the repère $(\mathfrak{o}, e_i)$ the scalar square of the vector $u = p - \mathfrak{o}$ can be put in the form

$$g(u, u) = c^2t^2 - x^2 - y^2 - z^2,$$

where

$$p = xe_1 + ye_2 + ze_3 + te_4 + \mathfrak{o}.$$

Let us return to Galilean spacetime. Introducing scalar product into it becomes possible if we assume the existence of ether. Recall that Galilean spacetime was defined as a four-dimensional affine space with two metric elements: an absolute time form $\tau \in V^*$ and a scalar product h in the subspace $S = \ker \tau$ of spatial vectors. This model of spacetime was sufficient for the description of mechanical phenomena. To describe electromagnetic phenomena in it, it was necessary to bring in the ether, a non-spatial vector e. We adopt the convention that this vector is normalized, i.e. $\tau(e) = 1$. For any vector $k \in V$ we can write

$$k = e\tau(k) + (k - e\tau(k)).$$

It is easy to see that $\tau(k - e\tau(k)) = 0$, so that $k - e\tau(k)$ is a spatial vector (belongs to S) and therefore its scalar square is defined. If $k \in V$ is a tangent vector to the world-line of a light ray (the propagation vector), then the equality

$$\frac{\sqrt{h(k - e\tau(k), k - e\tau(k))}}{\tau(k)} = c$$

is a mathematical expression of the fact that, relative to the ether, light propagates at a constant velocity, c.

We can also proceed the other way round. Namely, define a map g: $V \times V \to \mathbf{R}^1$ as follows ($u, v \in V$):

$$g(u, v) = c^2\tau(u)\tau(v) - h(u - e\tau(u), v - e\tau(v)).$$

The equality $g(k, k) = 0$ occurs when, and only when, k is a directional vector of a light ray. Vector satisfying the equation $g(k, k) = 0$ are called *null vectors*. Thus null vectors are the vectors of light propagation (Fig. 6.2). For this reason, the term "null vector" is often replaced by "light vector".

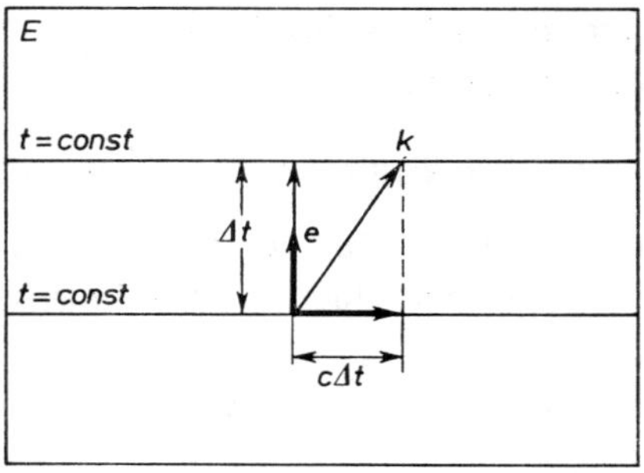

Fig. 6.2

The mapping g possesses all the properties of a scalar product: it is bilinear and symmetric. Furthermore, it is non-singular, i.e. if $g(u, v) = 0$ for every vector $v \in V$, then $u = 0$. Indeed, by substituting $v = e$, we obtain $g(u,e) = c^2\tau(u) = 0$, so $u \in S$. In S, however, g is identical with the scalar product $-h$, which is non-singular. We have thus shown that the only vector perpendicular to all vectors in V, in the sense of the scalar product g, is 0. The essential difference between g and h is that g is not positive-definite (or negative-definite), i.e. the equality $g(u, u) = 0$ does not imply $u = 0$.

We thus have the following geometric elements in spacetime E: τ, h and e. Alternatively, it is sufficient to give τ and g. Indeed, we can then define e

as a vector satisfying the conditions $\tau(e) = 1$ and $g(e, e) = c^2$, and define h by transforming the identity defining g. Similarly, an equivalent set of information about spacetime is provided by e and g. This spacetime, which should bear the names of Maxwell and Lorentz, constitutes a model for the description of mechanical and electromagnetic phenomena. To describe mechanical phenomena in this model, τ and h are sufficient. The scalar product g, on the other hand, turns out to be sufficient for the formulation of vacuum electrodynamics.

The step made by Einstein was that, of all these geometric elements, he only left the scalar product g in his model of spacetime. Einstein's theory, called the *special theory of relativity*, is the description of the properties of a spacetime which is a good "background" for all physical phenomena except those related to gravitation.

We say that the scalar product g has the signature $(+, -, -, -)$ if there is a basis (e_0, e_1, e_2, e_3) (the basis vector e_4 has been replaced by the vector $e_0 = e_4/c$) such that the matrix $g_{ij} = g(e_i, e_j)$ becomes

$$g_{ij} = \begin{cases} 1, & i = j = 0, \\ -1, & i = j = 1, 2, 3, \\ 0, & i \neq j. \end{cases}$$

A basis having this property is called *orthonormal.*

Let us express vectors $u, v \in V$ in this basis: $u = u^i e_i, v = v^i e_i$. The scalar product of u and v can then be written

$$g(u, v) = g_{ij} u^i v^j = u^0 \cdot v^0 - \mathbf{u} \cdot \mathbf{v},$$

where $\mathbf{u}$ and $\mathbf{v}$ are defined by $(u^i) = (u^0, \mathbf{u})$ and $(v^i) = (v^0, \mathbf{v})$, and the symbol $\mathbf{u} \cdot \mathbf{v}$ denotes the ordinary scalar product in three-dimensional Euclidean space $\mathbf{R}^3$.

Expressed relative to the *orthonormal repère* $(\mathfrak{o}, e_i)$, an event p can be written

$$p = cte_0 + xe_1 + ye_2 + ze_3 + \mathfrak{o}.$$

The zero coordinate has the form ct; here t is time with respect to the given orthonormal repère $(\mathfrak{o}, e_i)$.

Minkowski space (affine) is a three-tuple (E, V, g), where (E, V) is a four-dimensional affine space and g a scalar product in V with signature $(+, -, -, -)$. The pair (V, g) is called *Minkowski vector space.*

Minkowski space is the model of the special theory of relativity. This means that, from the mathematical point of view, the three-tuple (E, V, g) is the basis for almost all that comes under the name of special relativity. Why we say "almost" will become clear later.

Given a mathematical structure, one should study its automorphisms. Before, we investigated automorphisms of Galilean spacetime, called Galilean transformations. Now we shall consider automorphisms of Minkowski space-time, called Poincaré transformations.

A *Poincaré transformation* is a pair of bijections $f: E \to E$ and $\lambda_f: V \to V$, such that

(1) $f(p+u) = f(p) + \lambda_f(u)$ for any $u \in V$, $p \in E$,

(2) λ_f is a linear mapping,

(3) $g(\lambda_f u, \lambda_f v) = g(u, v)$, $(v \in V)$.

λ_f is determined by f (as suggested by notation) and is called a *Lorentz transformation.* Often an alternative terminology is used, in which f is called a Lorentz transformation and λ_f a homogeneous Lorentz transformation.

Poincaré transformations are automorphisms of Minkowski affine space, while Lorentz transformations are automorphisms of Minkowski vector space. Poincaré transformations constitute a group, called the *Poincaré group.* We shall denote it by $\mathrm{Aut}(E, V, g)$. Similarly, we have the Lorentz group, $\mathrm{Aut}(V, g)$. It is readily shown that $\lambda_{f_1 \circ f_2} = \lambda_{f_1} \circ \lambda_{f_2}$, so that the mapping $\lambda: \mathrm{Aut}(E, V, g) \to \mathrm{Aut}(V, g)$ is a homomorhpism of the Poincaré group onto the Lorentz group.

An example of Poincaré transformation is provided by translation. With a vector $v \in V$ we associate a mapping $\varkappa_v: E \to E$ such that $\varkappa_v(p) = v+p$ for any point $p \in E$. It is easy to see that $\varkappa_v$ is a Poincaré transformation such that the corresponding Lorentz transformation is the identity map in V. We note furthermore that $\varkappa_{v_1+v_2} = \varkappa_{v_1} \circ \varkappa_{v_2}$, so the mapping $\varkappa: V \to \mathrm{Aut}(E, V, g)$ is a (one-to-one) group homomorphism, if V is treated as a group with respect to addition. We thus obtain a sequence of homomorphisms:

$$0 \to V \xrightarrow{\varkappa} \mathrm{Aut}(E, V, g) \xrightarrow{\lambda} \mathrm{Aut}(V, g) \to 1\,.$$

This sequence is exact, i.e. the image of the homomorphism $\varkappa$ coincides with the kernel of the homomorphism λ. It turns out that the Poincaré group is a semi-simple product of the vector group V and the Lorentz group. The Poincaré group is the subject of intensive study by physicists, the reason being that almost all elementary particle kinematics can be reduced to the investigation of the properties of this group.

Vectors in V can be classified according to the sign of their scalar product. If

$$g(u, u) \begin{cases} > 0, & u \text{ is called a } \textit{timelike vector}, \\ = 0, & u \text{ is called a } \textit{null vector}, \\ < 0, & u \text{ is called a } \textit{spacelike vector}. \end{cases}$$

On this basis we classify other geometrical objects in Minkowski space. We say, for example, that two events p and q are timelike (or null, or spacelike) relative to one another if the vector $p-q$ is timelike (null, spacelike). All null events relative to an event $p \in E$ form a cone, called the *null* (or light) *cone* of p, which divides the remaining part of spacetime into three disjoint

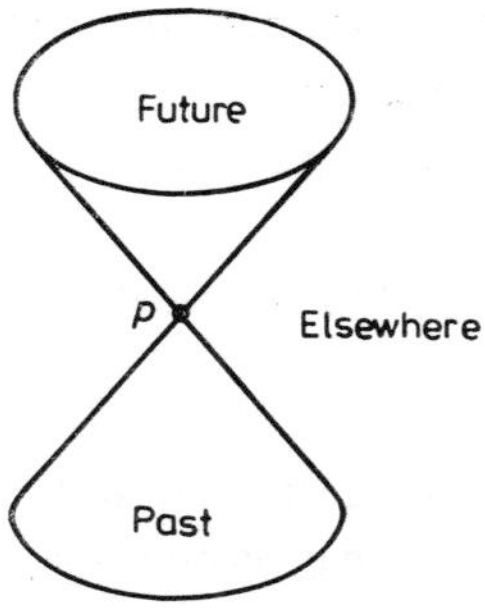

Fig. 6.3

regions (Fig. 6.3). Two of them contain timelike events relative to p, and one, spacelike events. The timelike regions are the interiors of the two sheets of the null cone. We call one of these regions the future and the other the past of the event p; we also include in the future and in the past the corresponding sheets of the cone.

Events which belong to the future of the event p (and only these events) can be reached from p by moving with a velocity not exceeding the velocity of light. The event p can only be reached from its past, if we restrict ourselves to motions with velocities less than the speed of light.

Events lying outside the null cone of the event p (i.e. events spacelike relative to p) do not stand in any causal relationship with p if only velocities less than c are considered, so we call this region "elsewhere".

The separation of the future from the past is a new element in our mathematical model of special relativity. We now have the complete model: the *time-oriented Minkowski space.* Temporal orientation enters into the fundamental laws of physics in a more discrete way than the scalar product g. More specifically, the temporal orientation does not intervene in the fundamental equations of physics themselves but only in the boundary conditions. Its intervention in electrodynamics, for example, manifests itself in that we accept

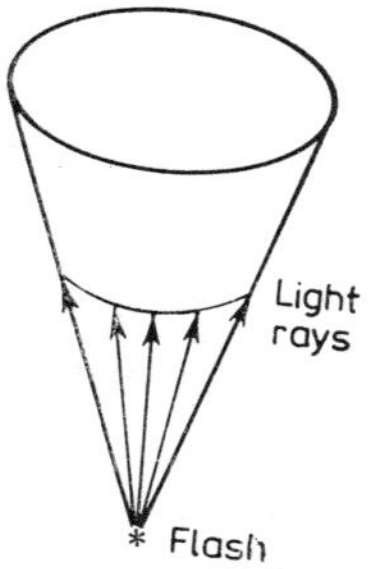

Fig. 6.4

the so-called retarded solutions of wave equations and reject the advanced solutions. In other words, we only allow the forward propogation of light in time, as shown in Fig. 6.4.

Among curves in E we distinguish *timelike curves, null curves* and *spacelike curves*. A timelike curve is a curve whose tangent vector is timelike at every point; null and spacetime curves are defined similarly. Analogous distinctions can be introduced for submanifolds of E with dimensions higher than 1.

As the first law of dynamics tells us, the world-lines of inertial observers are straight lines in E. The question arises whether they can be arbitrary lines.

Suppose that the world-line of an observer O is spacelike. Then there exists an event p such that no event on this world-line lies in the future of p; similarly, there is q such that the world-line of O does not pass through the past of q (Fig. 6.5). Observer O has no possibility of sending a light signal to q and, worse

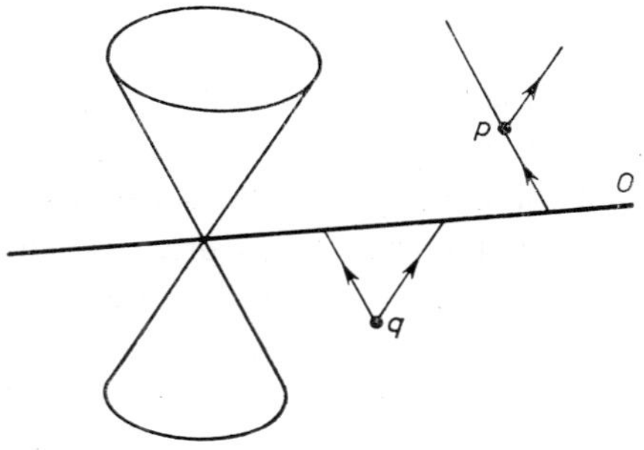

Fig. 6.5

still, he will never find out what happened at p (assuming he only uses light signals, or other signals with a speed not exceeding c).

If we agree that, in the end, all information about the external world is received through electromagnetic signals, we may assert that the world is not cognizable for the observer O, which appears to contradict experiment. Hence we conclude that there are no inertial observers with spacelike world-lines. Similarly we come to the conclusion that the world-lines of inertial observers cannot be null straight lines. It follows that the relative velocity of two inertial observers must be less than c.

The introduction of time orientation in spacetime establishes a temporal order among timelike and null events. No such order is possible among spacelike events. Every observer can tell whether an event p is earlier, than, simultaneous with, or later than another event q. But if p and q are spacelike relative to each other, there will always be some other observer who will see these events in a different temporal relationship.

Are motions with spacelike world-lines, i.e. motions with velocities greater than c relative to an inertial observer, possible? The theory of relativity does not give a categorical negative answer to this question.

Consider a light source S enclosed in a spherical shield of radius r with an aperture drilled in it. Let the shield rotate and the light beam fall on a spherical screen of radius R (Fig. 6.6). The angular velocity of the light spot on the screen

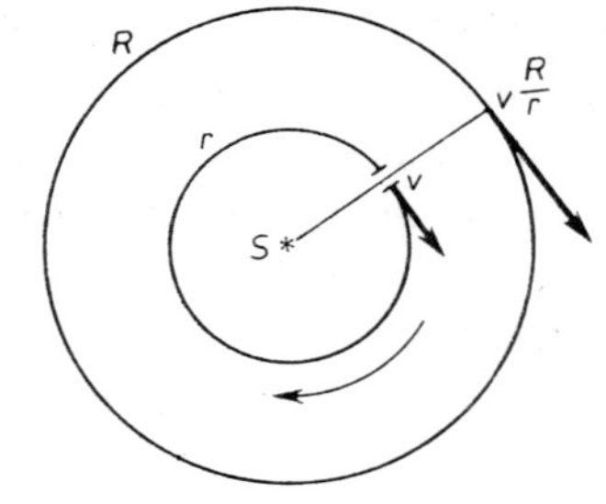

Fig. 6.6

is equal to the angular velocity of the shield a moment earlier. The ratio of the linear velocities of the light spot and the shield will therefore be R/r. Thus the velocity of the light spot on the screen can be arbitrarily large. This fact does not present any theoretical difficulty since the motion of the light spot does not involve transmission of energy or information.

Most physicists believe that *tachions*—hypothetical particles moving faster than light—do not exist. The existence of such particles, carrying energy or information, would lead to conclusions contradicting the principle of causality.

Imagine that an observer O sends a tachion which moves with a velocity $v > c$ and that this tachion hits an observer O' moving with a speed $V > c^2/v$. Consider two events p and q on the tachion's world-line, with p earlier than q from the point of view of O. To observer O', the event p will occur later than q. Thus no cause-effect relationship independent of the choice of observer can be identified between p and q. In particular, this applies to the transmission and reception of the tachion. Observer O' will assert that it was not him who was hit by the tachion but, on the contrary, that it was he who sent it. The assumption that the tachion could carry energy or information leads us to paradoxical conclusions.

We shall assume that transmission of energy and information with velocities greater than c is not possible. From this it follows immediately that there

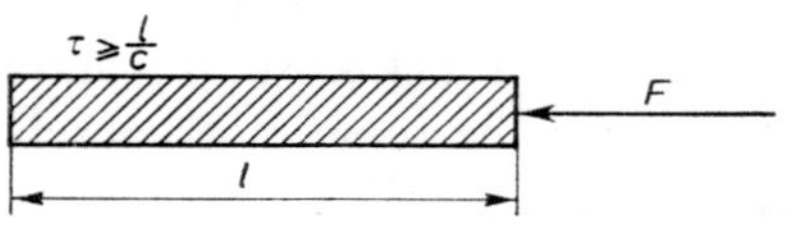

Fig. 6.7

are no bodies ideally rigid. For suppose that a force is applied to one end of a "rigid" rod with a rest length l. This end will move with an appropriate acceleration, whilst the other end will continue to be at rest for a period of time τ, not less than l/c, necessary for the information about what happened to the first end to reach it (Fig. 6.7). Thus we can only speak of rigid bodies when no forces act upon them, that is when they are in uniform and rectilinear motion. Hence the difficulties with rigid measuring rods in the early formulations of the theory of relativity.

The question of rigidity comes up in a certain well-known "paradox". Imagine a rod of rest length l_0 moving on a horizontal plank towards a hole with the same rest length. Suppose that the velocity of the rod is large enough for the relativistic length contraction to be significant, for example let $\sqrt{1-\beta^2} = 1/10$. A cursory conclusion from the contraction formula may be nonsensical: from the point of view of the plank the rod should fall into the hole, while from the point of view of the rod, it will not fit into it (Fig. 6.8).

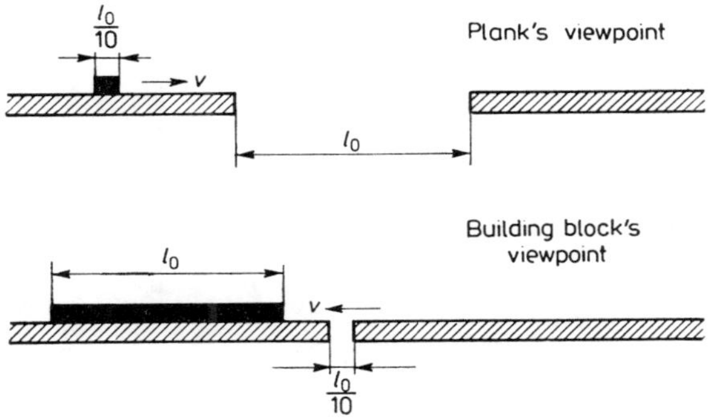

Fig. 6.8

In the light of our critical reappraisal of the concept of rigidity, the explanation of this "paradox" is not difficult. The immediate cause of the rod's falling into the hole is the force of gravity, and since a force is present the concept of rigidity no longer makes sense.

It is difficult to give an unequivocal answer to the question what will happen to the rod. The formulation of the problem leaves considerable freedom in the choice of parameters. The thickness of the plank and of the rod and their elastic properties will play a significant role even at low velocities. To describe the problem mathematically, we would first have to develop a mathematical model of elasticity in accordance with the theory of relativity. One may suppose that even a very long rod will bend, irrespective of velocity, and fall into the hole or strike its far edge, if both the rod and the plank are sufficiently thin.

That the description of the rod's falling into the hole is possible from the point of view of different reference frames can be verified by using the following idealization of the rod's motion. Assume that the hole has a cover which is removed—all its points simultaneously in the plank's reference frame—the instant the back end of the rod reaches it. Then the motion of every point of the rod in the plank's frame will be an ordinary projectile motion in a homogeneous gravitational field, and the rod will maintain its shape in this frame. In the frame in which it was initially at rest, however, the rod will be bent [41]. We encourage the reader to carry out an appropriate Lorentz transformation and give a description of the rod's motion in this frame.

The history of a material point is a curve in E. When E was Galilean spacetime, we parametrized such curves using the absolute time t. Now E is Minkowski spacetime, and we have no absolute time but times relative to arbitrary orthonormal repères. Let us recall that an event $p \in E$ in the history of a material point can be written relative to a repère $(\mathfrak{o}, e_i)$ as

$$p = x^i(t)e_i + \mathfrak{o},$$

where $x^0 = ct$, $x^1 = x(t)$, $x^2 = y(t)$, $x^3 = z(t)$ (Fig. 6.9). We want parametr-

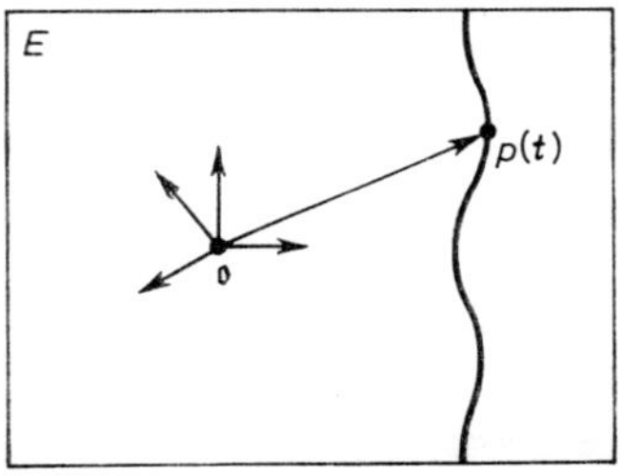

Fig. 6.9

ization independent of the choice of repère but determined solely by the geometrical elements of special relativity spacetime. An appropriate parameter is provided by the arc length s measured along the world-line by means of the scalar product g. Let us find the relation between the length s and the time t relative to repère $(\mathfrak{o}, e_i)$. In the basis (e_i), the tangent vector to the world-line has the form

$$\left(\frac{dx^i}{dt}\right) = \left(c, \frac{dx}{dt}, \frac{dy}{dt}, \frac{dz}{dt}\right) = (c, \mathbf{v}),$$

and its length is

$$\sqrt{c^2 - \left(\frac{dx}{dt}\right)^2 - \left(\frac{dy}{dt}\right)^2 - \left(\frac{dz}{dt}\right)^2} = c\sqrt{1-\beta^2}.$$

Hence the arc length as a function of t is given by

$$s(t) = \int_{t_0}^{t} c\sqrt{1 - \frac{v^2}{c^2}}\, dt',$$

which can also be written as

$$\frac{ds}{dt} = c\sqrt{1-\frac{v^2}{c^2}}.$$

The parameter s is called the *proper time*. The name derives from the fact that for $v = 0$ we have $s(t) = c(t-t_0)$, and therefore the proper time of a material point at rest relative to an inertial observer coincides (up to the constant factor c) with the time measured by the clock of that observer.

We can now give a definition of an ideal clock: An *ideal clock* is one which gives the proper time along its world-line irrespective of its motion. In Nature such clocks do not exist; we can only construct physical systems which approximate ideal clocks under given circumstances.

Looking back to the twin paradox, we can now describe it in a more realistic fashion: instead of approximating the world-line of the travelling twin by two straight line segments we shall leave it as a smooth curve (Fig. 6.10). Assuming

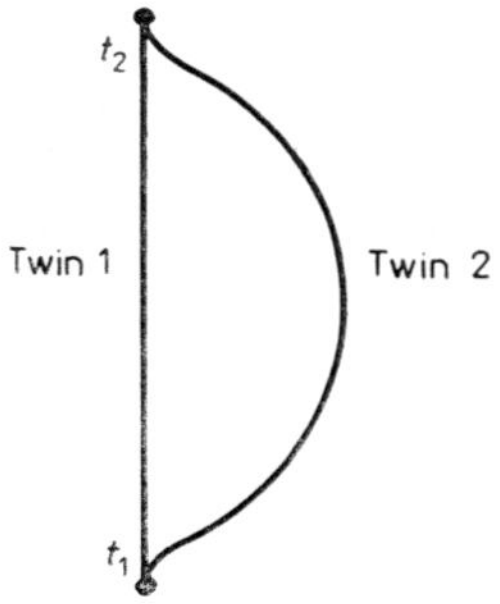

Fig. 6.10

that the biological clocks of the twins measure their proper times, we may conclude that the famous "paradox" follows from the inequality

$$\int_{t_1}^{t_2} c\sqrt{1-\frac{v^2}{c^2}}\,dt' \leqslant \int_{t_1}^{t_2} c\,dt',$$

where the equality occurs only when $v \equiv 0$. Of course the human "biological clock" is not ideal; the aging process depends on many factors, including the acceleration to which one is subjected, so the anecdote about the twins should not be taken too literally.

In employing the proper time, we act in accordance with Einstein's relativity principle, because we do not distinguish between inertial observers. The world-lines of material points will be parametrized, therefore, by the proper time s. In this parametrization, the tangent vector, which we shall call the *velocity four-vector*, or *four-velocity*, can be expressed in terms of the velocity relative to a given inertial reference frame as follows:

$$(u^i) = \left(\frac{\mathrm{d}x^i}{\mathrm{d}s}\right) = \left(\frac{1}{\sqrt{1-v^2/c^2}}, \frac{\mathbf{v}/c}{\sqrt{1-v^2/c^2}}\right).$$

Among tangent vectors, the four-velocity stands out as the one normalized to unity:

$$g(u, u) = g_{ij}u^i u^j = 1.$$

The counterpart of acceleration in special relativity is the *four-acceleration* w^i defined by

$$w^i = \mathrm{d}u^i/\mathrm{d}s.$$

This four-vector is orthogonal to the four-velocity, since by differentiating the identity $g_{ij}u^i u^j = 1$ we obtain $2g_{ij}u^i w^j = 0$. At velocities small compared with the speed of light, i.e. in a non-relativistic approximation, the four-acceleration reduces to

$$(w^i) \approx \left(0, \frac{\mathbf{a}}{c^2}\right).$$

How shall we define, in the relativistic framework, uniformly accelerated motion? The constancy of acceleration in any given inertial frame would lead, after sufficient time, to a velocity exceeding the velocity of light. The requirement that the acceleration four-vector be constant, on the other hand, cannot be reconciled with the unit norm of the four-velocity. We are forced to reject these two simplest possibilities. We note, instead, that at any instant of time the moving particle is at rest with respect to an observer whose world-line is tangent to the particle's world-line at that instant. In other words, at every instant of time the particle is at rest in some inertial frame. A uniformly accelerated motion will be defined as follows: it is a motion whose world-line is flat and whose acceleration relative to the instantaneous rest frame of the particle is constant.

Since the world-line is flat, there exists a rectilinear coordinate system such that it lies in the plane x^0, x^1. The velocity four-vector can then be written

$$u(s) = (\cosh\varphi(s), \sinh\varphi(s), 0, 0),$$

and the four-acceleration takes the form

$$w(s) = \frac{\mathrm{d}\varphi}{\mathrm{d}s}(\sinh\varphi, \cosh\varphi, 0, 0).$$

Relative to the instantaneous rest frame of the particle, the square of its acceleration,

$$a^2 = -c^4 g_{ij}w^i w^j|$$

should be constant. Thus, in uniformly accelerated motion, the four-acceleration has constant length. As a result, we obtain a differential equation for the function φ, from which we find

$$\varphi(s) = \frac{a}{c^2} s .$$

Integrating the equation $\frac{dx^i}{ds} = u^i$, after a convenient choice of coordinate origin we obtain the parametric equation of the world-line:

$$t(s) = \frac{c}{a} \sinh \frac{as}{c^2},$$

$$x(s) = \frac{c^2}{a} \cosh \frac{as}{c^2} .$$

We can see that

$$x^2 - c^2 t^2 = \frac{c^4}{a^2},$$

and so the world-line of uniformly accelerated motion is a hyperbola (Fig. 6.11).

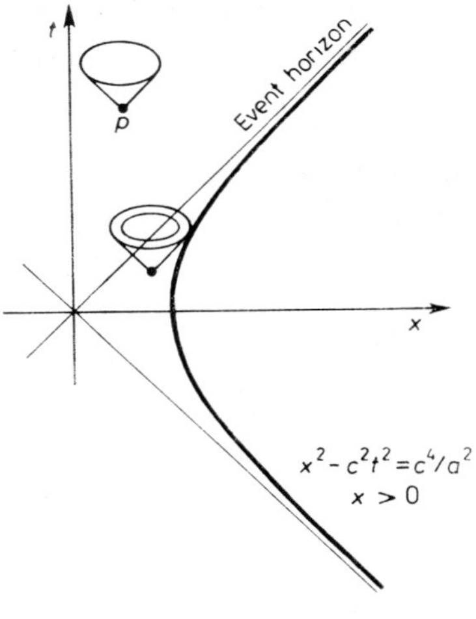

Fig. 6.11

In the chosen reference frame the trajectory of the motion is a straight line; in other inertial frames it will be a hyperbola, and not a parabola as in the non-relativistic case. The non-relativistic description approximates the relativistic one well if the duration of the motion is much less than c/a.

As we look at Fig. 6.11, we notice that an observer moving with uniform acceleration cannot send any information to the spacetime region defined by the inequality $ct \leqslant -x$. Neither can he receive information about events such as p, shown in the figure, for which $ct \geqslant x$. Because of this, we call the hyperplane $ct = x$ the event horizon of that observer.

In Galilean spacetime the four-velocity $u^i (u^i = \mathrm{d}x / \mathrm{d}t)$ satisfies the equation $\tau(u) = 1$. The velocity space $\{u \in V : \tau(u) = 1\}$ is a three-dimensional Euclidean space (Fig. 6.12). It is an affine space for which the vector space $S \subset V$ is the

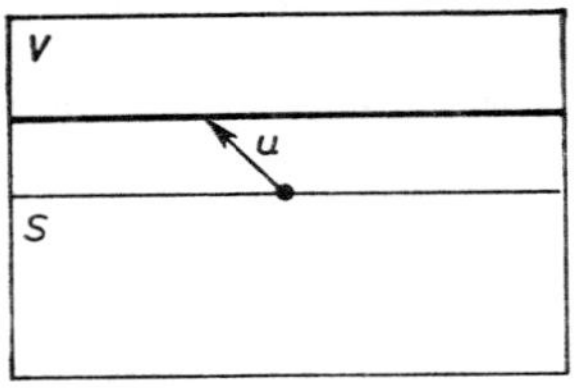

Fig. 6.12

space of translations. Owing to the existence of the scalar product h in S, a Euclidean distance is defined in it.

In Minkowski spacetime the situation is somewhat different. Here the four-velocity $u (u^i = \mathrm{d}x^i / \mathrm{d}s)$ satisfies the condition $g(u, u) = 1$, so that $u^0 = \pm\sqrt{1 + \mathbf{u}^2}$, and we have $u^0 \geqslant 1$ or $u^0 \leqslant -1$. Assuming, furthermore, that spacetime is temporally oriented and restricting ourselves to future-pointing velocity four-vectors, we can define the velocity space in special relativity as

$$\{u \in V : g(u, u) = 1 \text{ and } u^0 \geqslant 1\}.$$

This set is a three-dimensional Łobaczewski space, i.e. Riemannian manifold with constant negative curvature (Fig. 6.13). It can be shown that the distance between two points in this space is $\varrho(u_1, u_2) = \operatorname{arcosh} g(u_1, u_2)$. This formula

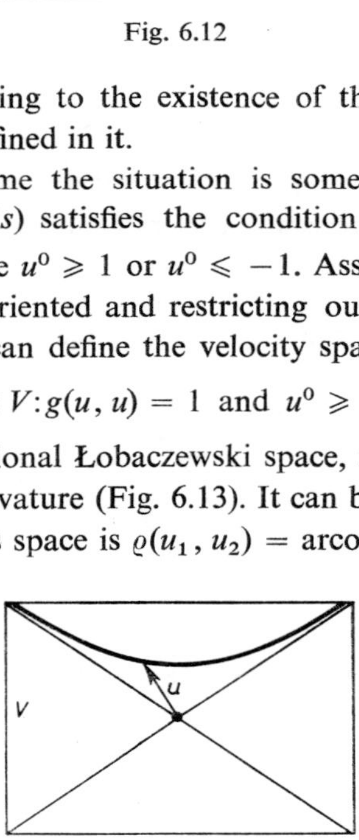

Fig. 6.13

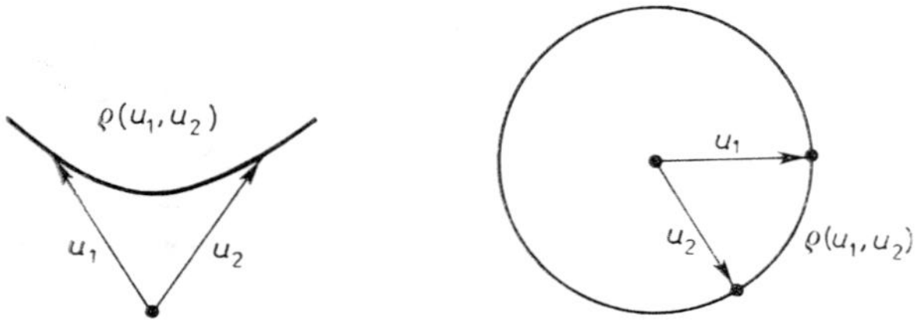

Fig. 6.14

is analogous to that for the distance between two points on the surface of a unit sphere, where we have $\varrho(\mathbf{u}_1, \mathbf{u}_2) = \arccos(\mathbf{u}_1 \cdot \mathbf{u}_2)$. The surface of a sphere is a manifold with a constant positive curvature (Fig. 6.14).

Imagine that an inertial observer O_1, whose world-line intersects the world-line of another observer O_2, sends a light signal to that observer (Fig. 6.15). The vector $\alpha u_2 - u_1$ is a directional vector of this signal, and therefore

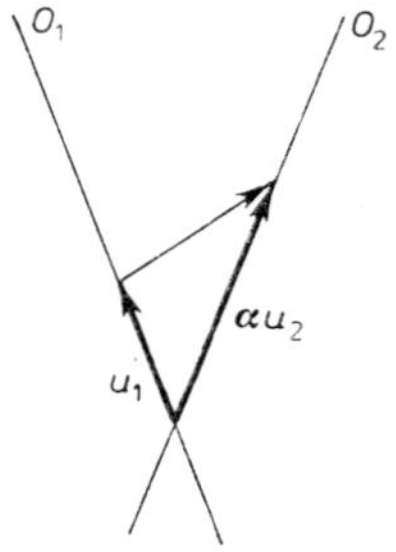

Fig. 6.15

a null vector. It follows that

$$\alpha^2 + 1 - 2\alpha g(u_1, u_2) = 0$$

or

$$g(u_1, u_2) = \frac{1}{2}\left(\alpha + \frac{1}{\alpha}\right).$$

Since $\alpha = \sqrt{(1+\beta)/(1-\beta)}$, we obtain

$$\varrho(u_1, u_2) = \operatorname{artanh}\beta.$$

If u_1, u_2 and u_3 are in one plane (Fig. 6.16), we have

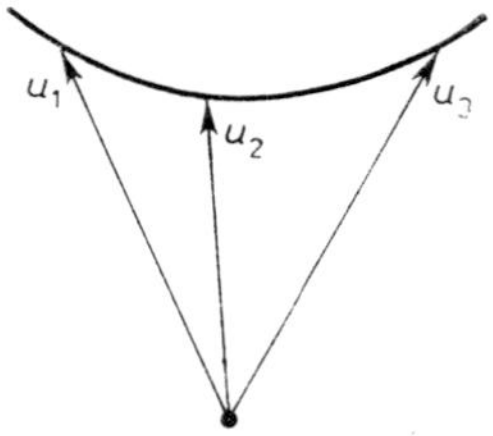

Fig. 6.16

$$\varrho(u_1, u_2) + \varrho(u_2, u_3) = \varrho(u_1, u_3),$$

(assuming $\varrho(u_1, u_3)$ is the largest of the three distances). Considering that

$$\tanh(x+y) = \frac{\tanh x + \tanh y}{1 + \tanh x \tanh y}$$

we obtain the known formula for the addition of velocities.

A large part of relativistic kinematics consists in applying the results of the geometry of Lobaczewski spaces. The above derivation of the velocity addition formula is one illustration of this application. It is interesting to see how optical phenomena come into play: light rays correspond to the absolute elements ("infinitely distant") in this geometry.

CHAPTER 7

The Lorentz Group and the Shape of Bodies in Motion

We shall now consider the geometry of optical observations. On this occasion, we shall analyse important properties of the Lorentz group. We shall consider idealized optical observations. We shall assume that they are made by an inertial observer O using a photographic plate having the shape of a sphere. This plate is exposed for a while at a time t. It bears the traces of events on the cone of the past, which contains the sphere at the moment of exposure and has its origin on the world-line of O. We should note that the recorded events are not as a rule simultaneous. The plate of an inertial observer O' moving with respect to O has its world-cylinder inclined relative to the latter (Fig. 7.1).

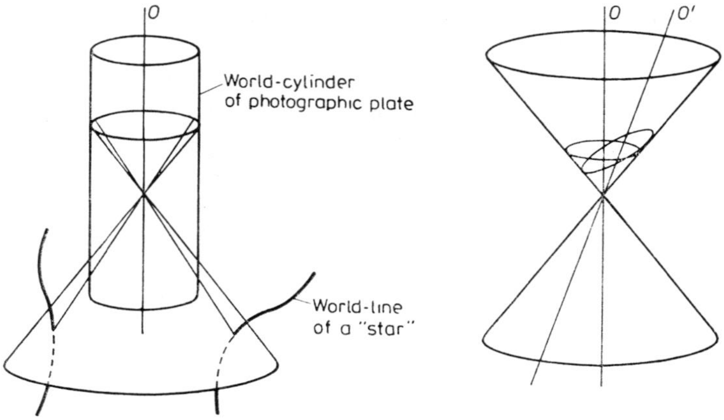

Fig. 7.1

We expose it at such a moment that the top of the cone in question is the point of intersection between the world-lines of O and O'. To analyse the relation between what has been recorded on O's plate and what will be registered on that of O' (Fig. 7.2), we must turn to mathematics.

Any basis $e = (e_i)$ in a Minkowski vector space determines an isomorphism $e\colon \mathbf{R}^4 \to V$, such that

$$e(u) = u^i e_i, \quad \text{where} \quad u = u^i \in \mathbf{R}^4.$$

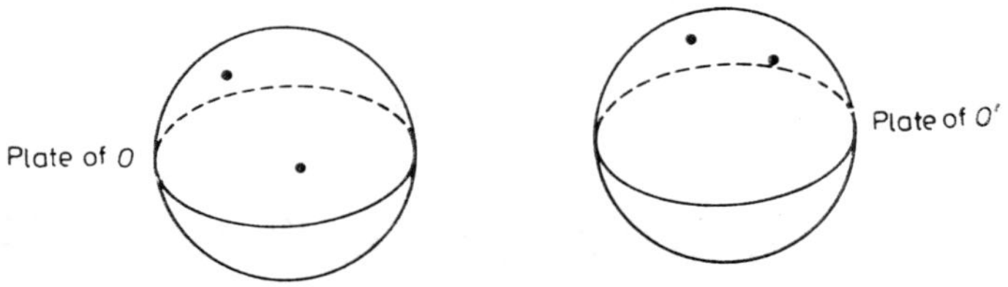

Fig. 7.2

If the basis e is orthonormal, the scalar product η in $\mathbf{R}^4$ induced by g, has the form (in this chapter $c = 1$)

$$\eta(v, v) = v^2 = t^2 - x^2 - y^2 - z^2,$$

where $v = (t, x, y, z)$. Later on, instead of Minkowski space (V, g), we shall use the space $\mathbf{R}^4$ with the scalar product η, bearing in mind that this procedure singles out a certain basis in V.

The plan of our further considerations is as follows. We shall try to find the different relations between various spaces and also those between the groups acting in them. It will be found that we can replace the relatively complex section of the Lorentz group in the set of light rays by the action of the homography group in the complex plane, which is easier to describe. Knowing the properties of the latter group, we can draw conclusions about the Lorentz transformations. First, we shall consider a stereographic projection (Fig. 7.3).

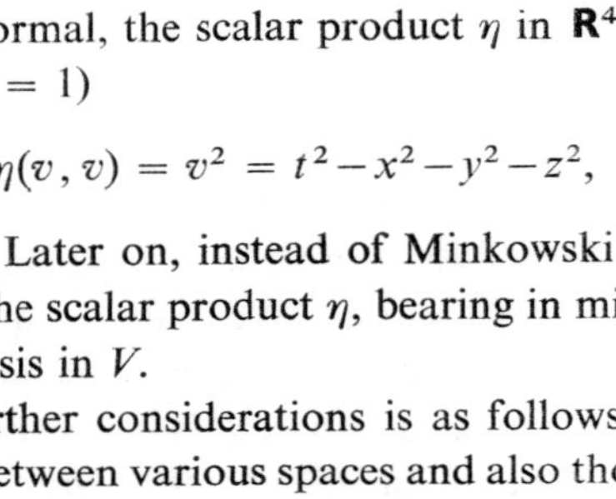

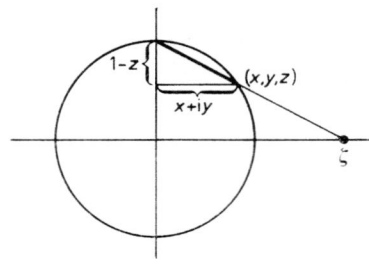

Fig. 7.3

It consists of mapping a sphere onto the complex plane complemented with a point "in infinity". We assign

$$\zeta = \frac{x+iy}{1-z}, \quad \text{if} \quad z = 1,$$

and

$$\zeta = \infty, \quad \text{if} \quad z = 1,$$

to a point on the sphere characterized by the numbers x, y, z which are related by the equality $x^2 + y^2 + z^2 = 1$.

The stereographic projection is a bijection. Using the equality $xdx+ydy+$ $+zdz = 0$ we have

$$d\zeta d\bar{\zeta} = \frac{dx^2+dy^2+dz^2}{(1-z)^2} .$$

We see that the metrics on the sphere and the complex plane are proportional to each other. We say that the stereographic projection is a conformal map.

Conformal maps are characterized by the fact that, although they do not preserve the distances (as they are not isometries), they preserve the angles. The angle between two curves on the sphere is equal to the angle between the projections of these curves onto the plane. The stereographic projection transforms circles on the sphere onto circles on the plane, or, if the circle crosses the point $x = 0$, $y = 0$, $z = 1$, onto a straight line.

We shall now discuss the relation between $\mathbf{R}^4$ with the scalar product η and the set $\mathbf{H}$ of second-degree Hermitian matrices. We define the mapping $\sigma: \mathbf{R}^4 \dashrightarrow \mathbf{H}$,

$$\sigma(t, y, y, z) = \begin{pmatrix} t+z & x-iy \\ x+iy & t-z \end{pmatrix}.$$

We can readily see that the image of any vector $v = (t, z, y, z)$ under the mapping σ is indeed a Hermitian matrix $\sigma(v) = \sigma(v)^\dagger$. The mapping σ is an isomorphism of vector spaces. An interesting property of this mapping is that the determinant of the matrix $\sigma(v)$ is equal to the scalar square of the vector v,

$$\det\sigma(v) = v^2.$$

In passing, we can note something about the basis in the vector space $\mathbf{H}$. Namely, the basis singled out by the mapping σ is the image of the canonical basis in $\mathbf{R}^4$:

$$\sigma(1, 0, 0, 0) = I,$$
$$\sigma(0, 1, 0, 0) = \sigma_x,$$
$$\sigma(0, 0, 1, 0) = \sigma_y,$$
$$\sigma(0, 0, 0, 1) = \sigma_z.$$

The matrices σ_x, σ_y, σ_z are called *Pauli matrices*. They were introduced to describe the electron spin.

Let $\mathbf{SL}(2, \mathbf{C})$ denote a group of unimodular (with the determinant equal to 1) complex matrices of the second degree. If $U \in \mathbf{SL}(2, \mathbf{C})$ and $A \in \mathbf{H}$, then also $\bar{U}AU^{\mathrm{T}} \in \mathbf{H}$, because

$$(\bar{U}AU^{\mathrm{T}})^\dagger = (U^{\mathrm{T}})^\dagger A^\dagger(\bar{U})^\dagger = \bar{U}AU^{\mathrm{T}}.$$

Moreover, the mapping $A \mapsto \bar{U}AU^{\mathrm{T}}$ is an automorphism of the vector space $\mathbf{H}$ and $\det(\bar{U}AU^{\mathrm{T}}) = \det A$. Just to ensure the latter equality, we require that the matrix U be unimodular (it is sufficient here that the module of the determinant U is equal to 1). Due to the fact that the transformation of **H** by means of the matrix U and the mapping σ are isomorphisms, there exists a unique real matrix $k(U)$ of the fourth degree, such that for each $v \in \mathbf{R}^4$ we have

$$\bar{U}\sigma(v)\,U^{\mathrm{T}} = \sigma\big(k(U)v\big).$$

Since U is unimodular, $k(U)$ preserves the scalar product in $\mathbf{R}^4$, therefore, $k(U)$ is an element of the Lorentz group $\mathbf{L} = \mathrm{Aut}(\mathbf{R}^4, \eta)$. The mapping

$$k\colon\ \mathbf{SL}(2, \mathbf{C}) \to \mathbf{L}$$

is a group homomorphism, as we have

$$\sigma\big(k(U_1 U_2)v\big) = \bar{U}_1 \bar{U}_2\, \sigma(v)(U_1 U_2)^{\mathrm{T}} = \bar{U}_1 \bar{U}_2\, \sigma(v)\, U_2^{\mathrm{T}} \cdot U_1^{\mathrm{T}} = \sigma\big(k(U_1)k(U_2)v\big).$$

In connection with the introduction of the homomorphism k we now make a few remarks about the topological properties of the Lorentz group.

Writing the vectors $u = (u^i) \in \mathbf{R}^4$ and $v = (v^i) \in \mathbf{R}^4$ in the form of columns, we can represent their scalar product as

$$\eta(u, v) = u^{\mathrm{T}}\eta v\,,$$

where η is the so-called *Minkowski metric matrix*

$$\eta = \begin{pmatrix} 1 & & & 0 \\ & -1 & & \\ & & -1 & \\ 0 & & & -1 \end{pmatrix},$$

(We hope that the fact that we denote them etric matrix and the scalar product by the same symbol does not cause any misunderstanding). The Lorentz group $\mathbf{L} = \mathrm{Aut}(\mathbf{R}^4, \eta)$ consists of real 4th-degree square matrices Λ satisfying the conditions

$$\Lambda^{\mathrm{T}}\eta\Lambda = \eta\,.$$

Such matrices are called *Lorentz matrices.* Taking the determinant of both sides of this equality, we obtain that $(\det\Lambda)^2 = 1$, therefore

$$\det\Lambda = 1 \quad \text{or} \quad \det\Lambda = -1\,.$$

The Lorentz group consists of two subsets: the subset $\mathbf{L}_+$, which consists of matrices with the determinant equal to $+1$, called *proper Lorentz matrices* (*transformations*), and the subset $\mathbf{L}_-$, containing improper Lorentz matrices, i.e., those with the determinant equal to -1. The subset $\mathbf{L}_+$ of the Lorentz group is its subgroup, called the proper Lorentz group. The proper Lorentz transformations preserve the orientation in Minkowski vector space.

The 00-component of the algebraic condition defining Lorentz matrices has the form

$$(\Lambda_{00})^2-(\Lambda_{10})^2-(\Lambda_{20})^2-(\Lambda_{30})^2 = 1,$$

therefore $(\Lambda_{00})^2 \geqslant 1$. We see thus that two cases can exist:

$$\Lambda_{00} \geqslant 1 \quad \text{or} \quad \Lambda_{00} \leqslant -1.$$

If $\Lambda_{00} \geqslant 1$, the Lorentz transformation corresponding to the matrix Λ preserves the time orientation, i.e. it transforms vectors directed to the future (timelike and null vectors) into those directed to the future, and vectors directed to the past into those directed to the past. If $\Lambda_{00} \leqslant -1$, Λ changes the time orientation of timelike and null vectors. Thus, the Lorentz group consists of the subsets $\mathbf{L}^{\uparrow}$ and $\mathbf{L}_{\downarrow}$ containing transformations which preserve the time orientation and those which change it respectively. The subset $\mathbf{L}^{\uparrow}$ of the Lorentz group is its subgroup, called the orthochronic Lorentz group. It is a group of automorphisms of Minkowski vector space with time orientation.

The Lorentz group is the disjoint sum of four components:

$\mathbf{L}_{+}^{\uparrow} = \mathbf{L}_{+} \cap \mathbf{L}^{\uparrow}$, *proper orthochronic,*

$\mathbf{L}_{+}^{\downarrow} = \mathbf{L}_{+} \cap \mathbf{L}^{\downarrow}$, *proper antichronic,*

$\mathbf{L}_{-}^{\uparrow} = \mathbf{L}_{-} \cap \mathbf{L}^{\uparrow}$, *improper orthochronic,*

$\mathbf{L}_{-}^{\downarrow} = \mathbf{L}_{-} \cap \mathbf{L}^{\downarrow}$, *improper antichronic.*

These four components are connected, which means that any two points belonging to one of the components may be connected by a curve which lies within this component. The connected component of unity $\mathbf{L}_{+}^{\uparrow}$ is a subgroup of the complete Lorentz group, called the *proper orthochronic Lorentz group.*

Since $\mathbf{SL}(2, \mathbf{C})$ is a connected group and the homomorphism k is continuous, the image of the group $\mathbf{SL}(2, \mathbf{C})$ under the mapping k must be contained in $\mathbf{L}_{+}^{\uparrow}$. To be exact, this image coincides with $\mathbf{L}_{+}^{\uparrow}$, $k(\mathbf{SL}(2, \mathbf{C})) = \mathbf{L}_{+}^{\uparrow}$. We can readily see that

$$k(-U) = k(U),$$

and we can also demonstrate that the question of sign is the only ambiguity in choosing U, if $k(U)$ is given. In this way, we can show that the sequence of homomorphisms

$$\mathbf{Z}_2 \underset{i}{\rightarrow} \mathbf{SL}(2, \mathbf{C}) \underset{k}{\rightarrow} \mathbf{L}_{+}^{\uparrow}$$

is exact. Here, we denote $\mathbf{Z}_2 = \{I, -I\}$, while i is the natural injection of $\mathbf{Z}_2$ into $\mathbf{SL}(2, \mathbf{C})$. In other words, $\mathbf{SL}(2, \mathbf{C})$ is an extension of the group $\mathbf{L}_{+}^{\uparrow}$. We can also show that the unimodular group is the minimal simple connected extension of the proper orthochronic Lorentz group.

We have discussed the isomorphism between $(\mathbf{R}^4, \eta)$ and $\mathbf{H}$, and the relation between groups acting in these spaces, namely $\mathbf{L}_{+}^{\uparrow}$ and $\mathbf{SL}(2, \mathbf{C})$.

We now introduce the notions of projective spaces and projective quadrics. Let V denote a vector space over the field of real numbers (without any greater modifications, we can apply this definition to vector spaces over the field of complex numbers). Let $0 \neq v \in V$; we will then call

$$d(v) = \{\lambda v : 0 \neq \lambda \in R\}$$

the direction of vector v, i.e., vectors which are proportional to one another have the same direction. The set of directions $P(V) = \{d(v) : 0 \neq v \in V\}$ is called the projective space determined by V. If a scalar product g is given in V, we can define the quadric $N(V)$, i.e. the set of null directions, in the following way:

$$P(V) \supset N(V) = \{d(v) : 0 \neq v \in V \text{ and } g(v, v) = 0\}.$$

If, in particular, we take $\mathbf{R}^4$ and η, we can form $N(\mathbf{R}^4)$, while taking $\mathbf{H}$ and the scalar product generated by the determinant, we can construct $N(\mathbf{H})$. The mapping σ determines the bijection $N(\mathbf{R}^4) \to N(\mathbf{H})$. Let us investigate the structure of the set $N(\mathbf{H})$. We have $d(A) \in N(\mathbf{H})$ if and only if $0 \neq A = A^\dagger$ and $\det A = 0$. Thus, we have two possibilities: either

$$A = \begin{pmatrix} \cdot & \cdot \\ \cdot & 0 \end{pmatrix}, \text{ and hence } A = \lambda \begin{pmatrix} 1 & 0 \\ 0 & 0 \end{pmatrix}, \text{ where } 0 \neq \lambda \in \mathbf{R},$$

or

$$A = \lambda \begin{pmatrix} \cdot & \cdot \\ \cdot & 1 \end{pmatrix}, \text{ and hence } A = \lambda \begin{pmatrix} \zeta\bar{\zeta} & \bar{\zeta} \\ \zeta & 1 \end{pmatrix}, \text{ where } 0 \neq \lambda \in \mathbf{R},\ \zeta \in \mathbf{C}.$$

When $\zeta \to \infty$, the matrix of the second type tends to that of the first type (to verify this, you need to choose $\lambda = \lambda'/\zeta\bar{\zeta}$, where $\lambda' = \text{const}$). We thus have a continuous bijection $N(\mathbf{H}) \to \tilde{\mathbf{C}} = \mathbf{C} \cup \{\infty\}$, described by

$$d\begin{pmatrix} 1 & 0 \\ 0 & 0 \end{pmatrix} \to \infty, \quad d\begin{pmatrix} \zeta\bar{\zeta} & \bar{\zeta} \\ \zeta & 1 \end{pmatrix} \to \zeta.$$

Let us, moreover, note that there exists a bijection of the two-dimensional sphere $\mathbf{S}_2$ onto the set of isotropic directions in $(\mathbf{R}^4, \eta)$: Namely, if $(x, y, z) \in \mathbf{S}_2$, i.e., if $x^2 + y^2 + z^2 = 1$, then $d(1, x, y, z) \in N(\mathbf{R}^4)$.

We can now compose the three mappings

$$\mathbf{S}_2 \to N(\mathbf{R}^4) \to N(\mathbf{H}) \to \tilde{\mathbf{C}},$$

according to the following rule

$$(x, y, z) \mapsto d(1, y, y, z) \mapsto d\begin{pmatrix} 1+z & x-\mathrm{i}y \\ x+\mathrm{i}y & 1-z \end{pmatrix} \mapsto \begin{cases} \dfrac{x+\mathrm{i}y}{1-z}, & z \neq 1, \\ \infty, & z = 1. \end{cases}$$

We can see that this composition is a stereographic projection.

Let us now consider the action of the group **SL**(2, **C**) in $N(\mathbf{H})$, and how it is transferred to $\tilde{\mathbf{C}}$. Let U $\in$ **SL**(2, **C**),

$$U = \begin{pmatrix} a & b \\ c & d \end{pmatrix}, \qquad ad-bc = 1$$

and

$$A = \begin{pmatrix} \zeta\bar{\zeta} & \bar{\zeta}' \\ \zeta & 1 \end{pmatrix},$$

i.e., $d(A) \in N(\mathbf{H})$. Let us note that we can represent the matrix A as the product of two vectors:

$$A = \begin{pmatrix} \zeta\bar{\zeta} & \bar{\zeta} \\ \zeta & 1 \end{pmatrix} = \begin{pmatrix} \bar{\zeta} \\ 1 \end{pmatrix} (\zeta, 1) = \begin{pmatrix} \bar{\zeta} \\ 1 \end{pmatrix} \begin{pmatrix} \zeta \\ 1 \end{pmatrix}^{\mathrm{T}}.$$

We can now factorize the action of the matrix U on A:

$$\bar{U}AU^{\mathrm{T}} = \bar{U}\begin{pmatrix} \bar{\zeta} \\ 1 \end{pmatrix} \left[U \begin{pmatrix} \zeta \\ 1 \end{pmatrix} \right]^{\mathrm{T}},$$

thus, U acts as follows:

$$U\begin{pmatrix} \zeta \\ 1 \end{pmatrix} = \begin{pmatrix} a\zeta+b \\ c\zeta+d \end{pmatrix} = (c\zeta+d)\begin{pmatrix} \dfrac{a\zeta+b}{c\,\zeta+d} \\ 1 \end{pmatrix}, \qquad \text{if} \qquad c\zeta+d \neq 0;$$

if, in turn, $c\zeta+d = 0$,

$$U\begin{pmatrix} \zeta \\ 1 \end{pmatrix} = \begin{pmatrix} a\zeta+b \\ 0 \end{pmatrix}.$$

We can thus see that the matrix U induces in $\tilde{\mathbf{C}}$ a mapping j_U called a *homography*,

$$j_U(\zeta) = \frac{a\zeta+b}{c\zeta+d}.$$

In this way, we have obtained the exact sequence of homomorphisms:

$$\mathbf{Z}_2 \xrightarrow[i]{} \mathbf{SL}(2, \mathbf{C}) \xrightarrow[j]{} \mathbf{PGL}(2, \mathbf{C}),$$

where **PGL**(2, **C**) denotes the *homography group* (i.e. the projective group). Hence, we can easily obtain the isomorphisms of the following groups:

$\mathbf{L}_+^\uparrow \leftrightarrow$ **PGL**(2, **C**) $\leftrightarrow$ conformal group of the sphere $\mathbf{S}_2$, and also the isomorphisms of spaces where these groups act (the spaces are homogeneous with respect to these groups):

$$N(\mathbf{R}^4) \leftrightarrow \tilde{\mathbf{C}} \leftrightarrow \mathbf{S}_2.$$

Homographies, being analytical mappings, are conformal on $\tilde{\mathbf{C}}$. Moreover, homographies preserve the set of straight lines and circles on the complex plane.

We can now answer the question: what is the relationship between the images on the plates of the observers O and O'? Since both observers are inertial, their celestial spheres $N(\mathbf{R}^4)$ and $N(\mathbf{R}^4)'$ are connected by the Lorentz transformation. These conformal transformations correspond to the $\mathbf{S}_2$ and $\mathbf{S}_2'$, obtained according to the previously given rule. We can thus conclude that the image of a circle on the plate is always a circle [38]. Therefore, it is not true that the wheels of a bicycle moving very fast relative to the observers look like ellipses [50, 58]. Such statements can be found in certain books popularizing the theory of relativity, e.g., in an otherwise splendid little book by Gamov, *Mister Tompkins in Wonderland* [23].

We have gained a convenient way of classifying Lorentz transformations: we can do it by classifying homographies. Let us consider how many directions on a cone can be preserved by a Lorentz transformation. This is equivalent to the question: how many solutions does the following equation have

$$\frac{a\zeta+b}{c\zeta+d} = \zeta\,?$$

After transformations we obtain the quadratic equation

$$c\zeta^2+(d-a)\zeta-b = 0,$$

whose discriminant

$$\Delta = (d-a)^2+4bc = (a+d)^2-4$$

defines the number of directions on the cone which are preserved. If $\Delta \neq 0$, two directions are preserved. If $\Delta = 0$, one direction is preserved, except in the case $b = c = 0$, $a = d = \pm 1$, when all the directions are preserved (the appropriate Lorentz transformation is then an identity).

The complexification of a real vector space. If V is a real vector space we can assign to it a complex vector space $V^{\mathbf{C}}$. Considered as the set $V^{\mathbf{C}}$, this space is the Cartesian product $V\times V$. Instead of writing $(u, v) \in V^{\mathbf{C}}$, it is convenient to use the notation $u+\mathrm{i}v \in V^{\mathbf{C}}$. The addition of vectors in $V^{\mathbf{C}}$ is then given by the formula

$$(u_1+\mathrm{i}v_1)+(u_2+\mathrm{i}v_2) = u_1+u_2+\mathrm{i}(v_i+v_2),$$

while multiplication by complex numbers is given by

$$(a+\mathrm{i}b)(u+\mathrm{i}v) = au-bv+\mathrm{i}(av+bu).$$

The complex dimension of the space $V^{\mathbf{C}}$ is equal to the real dimension of the space V.

The space $V^{\mathbf{C}}$ has a richer structure than an ordinary complex vector space, since it allows the definition of real vectors, i.e., vectors of the form $u+\mathrm{i}0$, imaginary vectors of the form $0+\mathrm{i}v$, and the complex conjugation: $\overline{u+\mathrm{i}v} = u-\mathrm{i}v$.

We can apply the above complexification procedure to the Minkowski vector space V. We can generalize the action of the scalar product in this space to cover complex vectors, according to the formula

$$g(u_1+\mathrm{i}v_1,\ u_2+\mathrm{i}v_2) = g(u_1, u_2)-g(v_1, v_2)+\mathrm{i}(g(u_1, v_2)+g(v_1, u_2)).$$

The space $V^{\mathbf{C}}$ thus obtained, with the scalar product g, is called the *complex Minkowski vector space.* The fact that the same letter is used to denote the scalar products both in space V and $V^{\mathbf{C}}$ should not cause any misunderstanding.

Having the Lorentz transformation $\varphi: V \to V$, we can determine the extension of this transformation to the complex Minkowski vector space $V^{\mathbf{C}}$ denoted by the same letter:

$$\varphi(u+\mathrm{i}v) = \varphi(u)+\mathrm{i}\varphi(v).$$

The transformation $\varphi: V^{\mathbf{C}} \to V^{\mathbf{C}}$ thus defined preserves the scalar product in the space $V^{\mathbf{C}}$.

In Minkowski space, we can introduce a real orthonormal basis (e_0, e_1, e_2, e_3). Frequently though, it is more convenient to use the null basis $(k, l, m, \bar{m})$, where the two vectors k and l are real and satisfy

$$g(k, l) = 1.$$

The other two vectors m and $\bar{m}$ are complex, mutually conjugate, and satisfy

$$g(m, \bar{m}) = -1,$$

while the remaining scalar products of vectors of this basis are equal to zero. We can find that such a basis exists by constructing it from an orthonormal basis in the following way:

$$k = \frac{e_0+e_3}{\sqrt{2}},$$

$$l = \frac{e_0-e_3}{\sqrt{2}},$$

$$m = \frac{e_1+\mathrm{i}e_2}{\sqrt{2}},$$

$$\bar{m} = \frac{e_1-\mathrm{i}e_2}{\sqrt{2}}.$$

The null basis is particularly convenient for the analysis of Lorentz transformations. We already know that the Lorentz transformation preserves at least one direction on the cone. We can choose the vector k in such a way that it points in this direction. Considering only those Lorentz transformations which preserve the time orientation, we find that the coefficient of proportionality between the vectors $\varphi(k)$ and k must be positive; therefore,

$$\varphi(k) = \mathrm{e}^{\psi}k.$$

Since the scalar product of the vectors $\varphi(k)$ and $\varphi(m)$ should vanish, there is no term proportional to the vector l in the decomposition

$$\varphi(m) = \alpha m + \beta \bar{m} + \gamma k.$$

Because $\varphi(m)$ should be a null vector, we have $\alpha\beta = 0$. Ignoring in these considerations transformations which do not preserve the orientation, we can eliminate the case $\alpha = 0$; therefore,

$$\beta = 0.$$

Next, we have

$$\varphi(\bar{m}) = \overline{\varphi(m)} = \bar{\alpha}\bar{m} + \bar{\gamma}k.$$

From the normalization condition $g(\varphi(m), \varphi(\bar{m})) = -1$, we obtain $\alpha\bar{\alpha} = 1$; therefore, we can write

$$\alpha = e^{i\varphi}.$$

It may now easily be found that the orthogonality conditions uniquely define the vector $\varphi(l)$ as

$$\varphi(l) = e^{-\psi}l + e^{-\psi}\gamma\bar{\gamma}k + \bar{\gamma}e^{-\psi+i\varphi}m + \gamma e^{-\psi-i\varphi}\bar{m}.$$

Through simple, though rather tedious, calculations we can find that the matrix $U \in \mathbf{SL}(2, \mathbf{C})$ corresponding to the above Lorentz transformation has the form

$$U = \pm \begin{pmatrix} e^{\frac{\psi}{2} - i\frac{\varphi}{2}} & \gamma e^{-\frac{\psi}{2} - i\frac{\varphi}{2}} \\ 0 & e^{-\frac{\psi}{2} + i\frac{\varphi}{2}} \end{pmatrix}.$$

We know that the number of directions on the cone which are preserved is determined by the quantity

$$\Delta = \left(e^{-\frac{\psi}{2} + i\frac{\varphi}{2}} - e^{\frac{\psi}{2} - i\frac{\varphi}{2}}\right)^2.$$

Thus, one direction is preserved if and only if

$$e^{\psi} = e^{i\varphi} = 1.$$

Let us consider first the case when two null directions are preserved. We can then further simplify the transformation formulae derived above, choosing the vector l so that it shows the other preserved direction. In such situation $\gamma = 0$. Therefore, the general form of the Lorentz transformation preserving two directions in this specially chosen basis is as follows:

$$\begin{aligned} \varphi(k) &= e^{\psi}k, \\ \varphi(l) &= e^{-\psi}l, \\ \varphi(m) &= e^{i\varphi}m. \end{aligned}$$

We can see that in-this case the Lorentz transformation consists of a special Lorentz transformation in the plane t, z (ψ is a hyperbolic angle) and rotation by an angle φ in the plane x, y which is perpendicular to the plane t, z. Transformations of this form make up a two-parameter Abelian subgroup of the Lorentz group.

Let us now consider a case where only one null direction is preserved. The general form of the Lorentz transformation (with a specially chosen vector k) is then as follows:

$$\begin{aligned}\varphi(k) &= k,\\ \varphi(l) &= l+\gamma\bar{\gamma}k+\bar{\gamma}m+\gamma\bar{m}\\ \varphi(m) &= m+\gamma k.\end{aligned}$$

Let us note that this Lorentz transformation preserves not only the chosen direction on the cone, but also the vectors pointing onto this direction. Let us note that transformations of this form also make up a two-parameter Abelian subgroup of the Lorentz group. It is interesting to observe that the above transformations are symmetries of plane electromagnetic waves.

We can write the matrix Λ of any Lorentz transformation in the form

$$\Lambda = e^{B},$$

where the matrix $B = (B^i_j)$, after lowering the upper index by means of g_{ij}, is antisymmetric, $B_{ij} = -B_{ji}$. We define the exponential function of the matrix in the above formula by means of the series

$$e^{B} = \sum_{n=0}^{\infty} \frac{B^n}{n!}.$$

This series is convergent for all matrices B.

In general, the above series contains an infinitely large number of terms. This is particularly so if we deal with a Lorentz transformation which preserves two null directions. In such a case, the matrix Λ in the basis $(k, l, m, \bar{m})$ has the form

$$\Lambda = \begin{pmatrix} e^{\psi} & 0 & 0 & 0\\ 0 & e^{-\psi} & 0 & 0\\ 0 & 0 & e^{i\varphi} & 0\\ 0 & 0 & 0 & e^{-i\varphi}\end{pmatrix},$$

while the corresponding matrix B is as follows:

$$B = \begin{pmatrix} \psi & 0 & 0 & 0\\ 0 & -\psi & 0 & 0\\ 0 & 0 & i\varphi & 0\\ 0 & 0 & 0 & -i\varphi\end{pmatrix}.$$

The case is different for a Lorentz transformation which preserves only one null direction. Here, the matrix

$$B = \begin{pmatrix} 0 & 0 & \gamma & \bar{\gamma} \\ 0 & 0 & 0 & 0 \\ 0 & \bar{\gamma} & 0 & 0 \\ 0 & \gamma & 0 & 0 \end{pmatrix},$$

whose third power vanishes, corresponds to the Lorentz matrix

$$\Lambda = \begin{pmatrix} 1 & \gamma\bar{\gamma} & \gamma & \bar{\gamma} \\ 0 & 1 & 0 & 0 \\ 0 & \bar{\gamma} & 1 & 0 \\ 0 & \gamma & 0 & 1 \end{pmatrix}.$$

Therefore, only the first three terms remain in the series e^B. Let us also remark that in this case we can write the matrix B in an invariant form

$$B^i{}_j = s^i k_j - s_j k^i.$$

k^i is the well-known null vector preserved by the Lorentz transformation under consideration, while s^i is a real spacelike vector, which is orthogonal to the vector k, and which can be defined by the formula

$$s = \bar{\gamma} m + \gamma \bar{m}.$$

The vector s is determined by the Lorentz matrix Λ to within the transformations $s \to s + \lambda k$, $\lambda \in \mathbf{R}$.

CHAPTER 8

Particles and Fields in Special Relativity Theory

The relativistic equation of the free motion of a material point is an equation of a straight line and therefore can be written

$$\frac{d^2x^i}{ds^2} = 0.$$

Since straight lines are curves for which the arc length attains an extremal value, the above equation is equivalent to the variational principle

$$\delta \int_{\text{world-line}} \text{const}\, ds = 0.$$

Choosing const $= -mc$ and using the relationship $ds = \sqrt{1-v^2/c^2}\,c\,dt$, we obtain the relativistic Lagrange function

$$L = -mc^2\sqrt{1-v^2/c^2} = -mc^2 + \frac{1}{2}mv^2 + O(1/c^2).$$

We can see that, to within a small of order $1/c^2$, this function differs from the non-relativistic Lagrange function $\frac{1}{2}mv^2$ only by a constant.

The generalized momentum corresponding to the relativistic Lagrange function is given by

$$\mathbf{p} = \frac{\partial L}{\partial \mathbf{v}} = \frac{m\mathbf{v}}{\sqrt{1-v^2/c^2}},$$

and the energy by

$$E = \mathbf{p}\cdot\mathbf{v} - L = \frac{mc^2}{\sqrt{1-v^2/c^2}}.$$

Together, the energy and the momentum form the *energy-momentum four-vector*

$$p^i = mcu^i,$$

the components of which are

$$p^0 = mc\frac{dx^0}{ds} = \frac{mc}{\sqrt{1-v^2/c^2}},$$

$$p^\alpha = mc\frac{dx^\alpha}{ds} = \frac{mv^\alpha}{\sqrt{1-v^2/c^2}}.$$

so that $(p^i) = (E/c, \mathbf{p})$.

It is now convenient to introduce the concept of *relativistic mass*,

$$m_{\text{rel}} = \frac{m}{\sqrt{1-v^2/c^2}},$$

for then the expression for momentum will take the form

$$\mathbf{p} = m_{\text{rel}}\mathbf{v}$$

analogous to the non-relativistic relationship between momentum and velocity. In terms of the relativistic mass, the energy is expressed by Einstein's famous formula

$$E = m_{\text{rel}}c^2.$$

The mass m is called the *rest mass*, since at zero velocity the relativistic mass is identical with m: $m_{\text{rel}}(v = 0) = m$.

The fact that the velocity four-vector is normalized implies that the energy-momentum four-vector is also normalized:

$$g_{ij}p^ip^j = m^2c^2;$$

hence we obtain the following relationship between energy and momentum

$$E^2 = m^2c^4+\mathbf{p}^2c^2.$$

Using the formula for the relativistic mass and the relationship between this mass and energy, we can express the energy E in a different form:

$$E = mc^2+\frac{1}{2}mv^2+O\left(\frac{1}{c^2}\right).$$

Thus, up to a small of order $1/c^2$, E differs from the energy $\frac{1}{2}mv^2$ of a free material point in Newtonian mechanics by the term mc^2. The quantity mc^2 is called the *rest energy*. Contrary to what may at first sight seem possible, the rest energy cannot be separated out from the total energy, for such a possibility would contradict the principle of relativity. Indeed, to carry out such a decomposition, one would have to select an inertial frame.

We will now present the equations of motion of a particle with charge q in an external electromagnetic field. This field is described by a *four-potential* $A = (A_i) = (\varphi, \mathbf{A})$. The quantity directly measured in experiments is the *electromagnetic field tensor*

$$F_{ij} = \partial A_j/\partial x^i - \partial A_i/\partial x^j.$$

The field tensor satisfies the *Maxwell equations*

$$\frac{\partial}{\partial x^j}F^{ij} = -\frac{4\pi}{c}j^i,$$

$$\frac{\partial}{\partial x^i} F_{jk} + \frac{\partial}{\partial x^j} F_{ki} + \frac{\partial}{\partial x^k} F_{ij} = 0 .$$

The form of these equations is identical in every inertial frame, in accordance with Einstein's principle of relativity. The first equation relates the field with the sources; j^i is the *current density four-vector*, $(j^i) = (c\varrho, \mathbf{j})$, where ϱ is the charge density and $\mathbf{j}$ the current density vector. The second equation ensures the existence of a four-potential for the field, but does not determine it uniquely. Adding the gradient of a function to the four-potential,

$$A_i \mapsto A_i + \partial \chi / \partial x^i,$$

does not change the field F_{ij}. This is the only freedom in the choice of four-potential.

Corresponding to the field F_{ij} in a given reference frame are the electric and magnetic field vector, $\mathbf{E} = (E_x, E_y, E_z)$ and $\mathbf{H} = (H_x, H_y, H_z)$, defined by

$$(F_{ij}) = \begin{pmatrix} 0 & E_y & E_y & E_z \\ -E_x & 0 & -H_z & H_y \\ -E_y & H_z & 0 & -H_x \\ -E_z & -H_y & H_x & 0 \end{pmatrix}.$$

In terms of the vectors $\mathbf{E}$ and $\mathbf{H}$, the Maxwell equations take the form

$$\operatorname{div} \mathbf{E} = 4\pi\varrho,$$

$$\operatorname{curl} \mathbf{H} - \frac{1}{c} \frac{\partial \mathbf{E}}{\partial t} = \frac{4\pi}{c} \mathbf{j} .$$

$$\operatorname{div} \mathbf{H} = 0,$$

$$\operatorname{curl} \mathbf{E} + \frac{1}{c} \frac{\partial \mathbf{H}}{\partial t} = 0 .$$

The metric tensor is used to lower indices, for example $u_j = g_{jk} u^k$. Let us introduce the contravariant metric tensor g^{jk}, defined by the equality

$$g_{ij} g^{jk} = \delta_i^k .$$

This tensor is used to raise indices, as in $F^{ij} = g^{ik} g^{jl} F_{kl}$.

The form of the equations of motion of a charged particle consistent with the principle of relativity is

$$\frac{\mathrm{d}p^i}{\mathrm{d}s} = \frac{q}{c} F^{ij} u_j .$$

Using the vectors $\mathbf{E}$ and $\mathbf{H}$, we can also write this equation as

$$\frac{\mathrm{d}\mathbf{p}}{\mathrm{d}t} = q\mathbf{E} + \frac{q}{c} \mathbf{v} \times \mathbf{H} .$$

The distribution of energy and momentum in a physical field or a continuous medium is described by means of the *energy-momentum tensor* T^{ij}. When integrated over a hypersurface Σ having a normal vector n_j, it yields the amount of energy and momentum flowing across Σ in the direction of the vector n_j:

$$p^i(\Sigma) = \frac{1}{c}\int_{\Sigma} T^{ij} n_j \, d\sigma .$$

Individual components of the energy-momentum tensor have the following meaning in a given reference frame:

T^{00}—energy density,

$\frac{1}{c}T^{\alpha 0}$—density of α-component of momentum,

$T^{0\alpha}$—α-component of energy flux,

$\frac{1}{c}T^{\alpha\beta}$—$\beta$-component of flux of α-component of momentum.

For example, $\frac{1}{c}T^{12}$ denotes the momentum component along the x-axis crossing a unit surface area perpendicular to the y-axis in unit time.

We will give two examples of the energy-momentum tensor. For a perfect fluid, it has the form

$$T^{ij} = (\varrho c^2 + p)u^i u^j - p g^{ij},$$

where u is the four-velocity of the fluid, ϱ its rest-frame density, and p its pressure. In the fluid's rest frame, in which $(u^i) = (1, 0, 0, 0)$, the components of the energy-momentum tensor become

$$\begin{aligned} T^{00} &= \varrho, \\ T^{0\alpha} &= T^{\alpha 0} = 0, \\ T^{\alpha\beta} &= p\delta^{\alpha\beta}. \end{aligned}$$

The energy-momentum tensor of an electromagnetic field is given by

$$4\pi T^{ij} = -F^{ik}F^{j}{}_{k} + \frac{1}{4}g^{ij}F_{kl}F^{kl}.$$

The law of conservation of energy-momentum asserts that the total inflow of energy-momentum into a four-dimensional region Ω is equal to zero:

$$\oint_{\partial\Omega} T^{ij} n_j \, d\sigma = 0,$$

where the normal vector n^j is oriented outward. In a special case, Ω may be the region enclosed between two hyperplanes of constant time in some inertial frame (Fig. 8.1). The hypersurfaces at spacelike infinity do not contribute

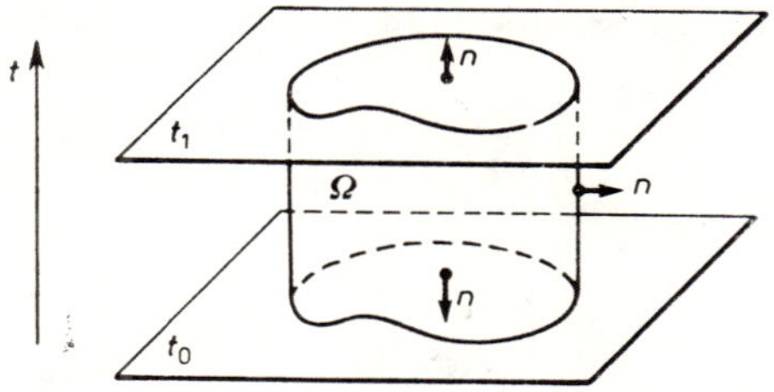

Fig. 8.1

to $\oint_{\partial\Omega} T^{ij}n_j\mathrm{d}\sigma$ if the energy-momentum tensor vanishes rapidly enough (faster than $1/r^2$) at large distances. Hence we obtain the global law of energy-momentum conservation

$$\int_{t=t_1} T^{i0}\mathrm{d}x\,\mathrm{d}y\,\mathrm{d}z = \int_{t=t_0} T^{i0}\mathrm{d}x\,\mathrm{d}y\,\mathrm{d}z.$$

The conservation law expressed by the previous formula is more general, since it also applies when the integrals over the timelike hypersurfaces of Fig. 8.1 do not tend to zero on passing to infinity, that is when radiation is present.

Converting the surface integral in the energy-momentum conservation law into a volume integral, we obtain

$$\int_{\Omega} \frac{\partial}{\partial x^j} T^{ij}\mathrm{d}\omega = 0 \qquad \text{for arbitrary region } \Omega,$$

and hence a differential formulation of the law of energy-momentum conservation:

$$\frac{\partial}{\partial x^j} T^{ij} = 0.$$

The *angular momentum density tensor* J^{ijk} is composed of an orbital part and a spin part:

$$J^{ijk} = \frac{1}{c}\,x^i T^{jk} - \frac{1}{c}\,x^j T^{ik} + S^{ijk}.$$

The *spin density tensor* S^{ijk}, like the tensor J^{ijk}, is antisymmetric in the first two indices: $S^{ijk} = -S^{jik}$.

The formula

$$J^{ij}(\Sigma) = \int_{\Sigma} J^{ijk} n_k \mathrm{d}\sigma$$

defines the amount of *angular momentum tensor* that crosses the hypersurface Σ in the direction of its normal vector. Like energy and momentum, angular momentum obeys the conservation law, which in an integral form reads

$$\oint_{\partial\Omega} J^{ijk} n_k \, d\sigma = 0,$$

and in a differential form

$$\frac{\partial}{\partial x^k} J^{ijk} = 0.$$

Combined with the differential law of energy-momentum conservation, the last equation gives us a formula for the antisymmetric part of the energy-momentum tensor

$$T^{ij} - T^{ji} = c \frac{\partial}{\partial x^k} S^{ijk}.$$

We can see that when the spin density vanishes, the energy-momentum tensor is symmetric. This happens for perfect fluids and for electromagnetic fields.

The angular momentum tensor of a body is defined by

$$J^{ij} = \int_{t=\mathrm{const}} J^{ij0} \, dx \, dy \, dz,$$

where integration extends over the region of space occupied by this body. If there is no ineraction between the body and the surrounding space, then the angular momentum tensor, and also the four-momentum of the body

$$p^i = \frac{1}{c} \int_{t=\mathrm{const}} T^{i0} \, dx \, dy \, dz$$

are conserved and hence independent of the hypersurface $t = \mathrm{const}$.

Let us note that under translation of the coordinate origin,

$$x'^i = x^i - a^i,$$

the four-momentum and the angular momentum tensor are transformed as follows

$$p'^i = p^i,$$
$$J'^{ij} = J^{ij} - a^i p^j + a^j p^i.$$

Suppose that we are given the momentum p^i and the angular momentum J^{ij}. Is it possible to determine the motion of the centre of mass, and so decompose the angular momentum into orbital and intrinsic parts by analogy with the nonrelativistic formula of rigid-body theory,

$$\mathbf{J} = \mathbf{R} \times \mathbf{p} + \hat{I} \boldsymbol{\omega},$$

where $\mathbf{R}$ is the centre-of-mass position vector, $\mathbf{p}$ the total momentum, $\boldsymbol{\omega}$ the angular velocity of the rigid body, and I the momentum of inertia tensor.

We shall call the system U' in which $J'^{ij} = 0$, the *centre-of-mass system.* According to the transformation law for the angular momentum tensor, the position vector X^i of the origin of this system should satisfy the equation

$$J^{ij}p_j - X^i p^j p_j + X^j p_j p^i = 0.$$

If $c^2m^2 = p^j p_j \neq 0$, the solution of this equation is

$$X^i = \frac{J^{ij}p_j}{c^2m^2} + \mu p^i,$$

where μ is an arbitrary real number.

In this way we have defined, for a body with non-zero rest mass, the world-line of the centre of mass. Its directional vector is the four-momentum. The angular momentum tensor in the centre-of-mass system is called the *spin tensor,* $S^{ij} = J'^{ij}$. We can now write the formula for the decomposition of the total angular momentum into the orbital and spin parts:

$$J^{ij} = X^i p^j - X^j p^i + S^{ij}.$$

Now suppose we demand that particles with zero rest mass, $p^j p_j = 0$, possess centre-of-mass systems, i.e. that the decomposition of the angular momentum tensor into orbital and spin parts be still valid for them, with $S^{ij}p_j = 0$. For particles of zero mass, the equation for the centre-of-mass position vector X^i takes the form

$$J^{ij}p_j + X^j p_j p^i = 0,$$

which can also be written as

$$J^{ij}p_j = \lambda p^i, \quad X^j p_j = -\lambda.$$

The first equation implies that the tensor J^{ij} is not arbitrary since it must have the null vector p^i as its eigenvector. The second equation asserts that the tip

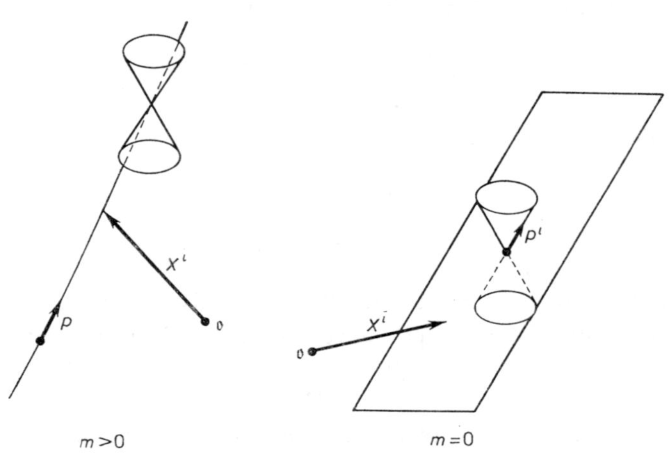

Fig. 8.2

of the vector X^j should lie on a hyperplane orthogonal to the vector p^j. Since p^j is a null vector, this means that the tip of X^j lies on a hyperplane tangent to the light cones—tangent along the straight lines for which p^j is the directional vector. This situation is illustrated in the right-hand side of Fig. 8.2; the left-hand side presents the analogous situation for particles with non-zero mass. From what has been said above it follows that particles of zero mass do not have well-defined world-lines and therefore cannot be located. Furthermore, their spin tensors are not uniquely defined since replacing vector X^i by a different "centre-of-mass" position vector in the decomposition of the angular momentum tensor into the orbital and spin parts will in general give a different value of S^{ij}.

Let ε_{ijkl} be the *Levi-Civitá pseudotensor*, i.e.

$$\varepsilon_{ijkl} = \begin{cases} +1 \text{ if } i, j, k, l \text{ form an even permutation of } 0, 1, 2, 3, \\ -1 \text{ if } i, j, k, l \text{ form an odd permutation of } 0, 1, 2, 3, \\ \;\;\,0 \text{ if two of the numbers } i, j, k, l \text{ are equal.} \end{cases}$$

Let us introduce the *Pauli-Lubański (spin) pseudovector*

$$S_l = \frac{1}{2}\varepsilon_{ijkl}p^i J^{jk} = \frac{1}{2}\varepsilon_{ijkl}p^i S^{jk}.$$

In the rest frame of a particle of mass $m > 0$, in which $(p^i) = (mc, 0, 0, 0)$, we have $(S_l) = mc(0, S_{23}, S_{31}, S_{12})$; hence the name. For particles of zero mass the pseudovector S_l is determined uniquely—it is independent of the choice of tensor S^{jk}. In both cases $S_l p^l = 0$.

Multiplying the above equality by ε^{lmpq}, we get

$$S_l \varepsilon^{lmpq} = p^m S^{pq} + p^p S^{qm} + p^q S^{mp}.$$

When the mass of the particle is non-zero, we obtain a formula expressing the spin tensor in terms of the spin pseudovector

$$S^{pq} = \frac{1}{m^2c^2} S_l p_m \varepsilon^{lmpq},$$

and when the mass vanishes, we get

$$S_l p_m \varepsilon^{lmpq} = 0,$$

or

$$S_l p_m - S_m p_l = 0.$$

It is not difficult to show that the fulfillment of this condition, which implies the proportionality of the spin pseudovector and the four-momentum

$$S_l = s p_l,$$

is equivalent to the possibility of decomposing (not uniquely) the angular momentum tensor of a massless particle into the orbital and spin parts.

The pseudoscalar s is called the *helicity* of the particle. For the particles of zero rest mass existing in Nature, it takes the values: $\pm\hbar$ for photons and $\hbar/2$ for neutrinos.

In every inertial reference frame the angular momentum tensor $J^{ij} = -J^{ij}$ defines two vectors: the *angular momentum vector* $\mathbf{J} = (J_{23}, J_{31}, J_{12})$, formed by the spatial components, and the *moment of energy vector* $\mathbf{K} = c(J_{01}, J_{02}, J_{03})$, formed by temporal-spatial components. The relativistic law of angular momentum conservation—the law of conservation of the tensor J^{ij}—contains two vector conservation laws. The law of conservation of the angular momentum vector is well known from classical mechanics; the meaning of the law of the moment of energy conservation is not so immediately clear. To interpret it from the point of view of classical mechanics, we should consider a body without spin, $S^{ij} = 0$. Then

$$\mathbf{K} = \mathbf{R}E - c^2 T\mathbf{p},$$

where $(cT, \mathbf{R}) = (X^i)$ is the centre-of-mass position four-vector. We can see that the law of conservation of moment of energy is equivalent to the law that the centre of (relativistic!) mass moves uniformly in a straight line. That the motion of the centre of mass is uniform and rectilinear, or, that the centre-of-mass world-line is a straight line, has already been established without the assumption of zero spin.

Are the energy-momentum and spin-density tensors determined uniquely by the distribution of matter? Consider the following transformation of the energy-momentum tensor:

$$T^{ij} \mapsto T'^{ij} = T^{ij} + \frac{\partial}{\partial x^k} A^{ijk},$$

where $A^{ijk} = -A^{ijk}$. Note that, under this transformation, T'^{ij} satisfies the law of energy-momentum conservation, as did T^{ij}. After the transformation, $p^i(\Sigma) \mapsto p'^i(\Sigma)$, where

$$p'^i(\Sigma) = \int_\Sigma T'^{ij} n_j \mathrm{d}\sigma = \int_\Sigma T^{ij} n_j \mathrm{d}\sigma + \int_\Sigma \frac{\partial}{\partial x^k} A^{ijk} n_j \mathrm{d}\sigma = p^i(\Sigma) + \oint_{\partial\Sigma} A^{ijk} \mathrm{d}\tau_{jk}.$$

The last integral has been converted from a hypersurface to a surface integral by using the four-dimensional Stokes theorem; $\mathrm{d}\tau_{jk}$ denotes the oriented surface element on $\partial\Sigma$.

If A^{ijk} vanishes on the boundary of the hypersurface Σ, then $p'^i(\Sigma) = p^i(\Sigma)$. If we let Σ be a hyperplane $t = \text{const}$, then $p'^i = p^i$ if A^{ijk} tends to zero faster than $1/r^2$ as $r \to \infty$. In this case, therefore, the tensors T^{ij} and T'^{ij} provide the same information about the global amount of energy and momentum. However, they localize energy and momentum differently, since for an arbitrary hypersurface Σ (which can also be a part of the hyperplane $t = \text{const}$) we will in general have $p'^i(\Sigma) \neq p^i(\Sigma)$. The equality $p'^i(\Sigma) = p^i(\Sigma)$ for every hypersurface Σ would imply $T'^{ij} = T^{ij}$.

Let us perform the following transformation of the spin density tensor together with the transformation of the energy-momentum tensor:

$$S^{ijk} \mapsto S'^{ijk} = S^{ijk} + \frac{1}{c} A^{jki} - \frac{1}{c} A^{ikj} + \frac{\partial}{\partial x^l} B^{ijkl},$$

where $B^{jikl} = -B^{ijkl}$, and $B^{ijlk} = -B^{ijkl}$. Then, the angular momentum density tensor, J^{ijk}, transforms to

$$J'^{ijk} = \frac{1}{c} x^i T'^{jk} - \frac{1}{c} x^j T'^{ik} + S'^{ijk} = J^{ijk} + \frac{\partial}{\partial x^l} C^{ijkl},$$

where

$$C^{ijkl} = \frac{1}{c} x^i A^{jkl} - \frac{1}{c} x^j A^{ikl} + B^{ijkl}.$$

Since $C^{jikl} = -C^{ijkl}$, the tensor J'^{ijk} satisfies the required property of antisymmetry in the first two indices. Since $C^{ijlk} = -C^{ijkl}$ also, we find that J'^{ijk} satisfies the conservation law, and therefore provides the same global description of angular momentum as J^{ijk} if the tensors A^{ijk} and B^{ijkl} die out rapidly enough as $r \to \infty$ — A^{ijk} faster than $1/r^3$, and B^{ijkl} faster than $1/r^2$.

The above transformations of the energy-momentum and spin density tensors allow the energy-momentum tensor to be symmetrized. To this end, ubstitute

$$A^{ijk} = \frac{c}{2}(S^{ijk} - S^{ikj} + S^{kji}),$$

$$B^{ijkl} = 0.$$

The transformed energy-momentum tensor

$$T'^{ij} = T^{ij} + \frac{c}{2} \frac{\partial}{\partial x^k} (S^{ijk} - S^{ikj} + S^{kji})$$

is symmetric, as follows from the formula for the antisymmetric part of the tensor T^{ij}, while the transformed spin density tensor vanishes:

$$S'^{ijk} = 0.$$

Let us note once more that the transformed energy-momentum and spin density tensors give the same global characteristics of energy-momentum and angular momentum as the original tensors; however,t hey localize these physical quantities differently.

CHAPTER 9

Spinors

It was P. A. M. Dirac who introduced spinors into relativistic physics, formulating the relativistic wave equation for particles with spin $\hbar/2$. In keeping with Dirac's postulates, this should be a first order equation, and after iteration, yield the Klein–Gordon equation

$$\left(\partial^i \partial_i + \frac{m^2c^2}{\hbar^2}\right)\psi = 0,$$

which is a quantum-mechanical counterpart of the relation $p^2 = m^2c^2$. Briefly, though not very accurately, we say that the Dirac equation is the *square root* of the Klein–Gordon equation. Earlier, through group theory considerations, E. Cartan had introduced spinors, while algebraic ideas related to spinors had their roots in works by Hamilton, Clifford and Lipschitz. Here, we will derive spinors algebraically.

Let us consider a real, or complex, n-dimensional vector space V with a symmetric, nondegenerate bilinear form (scalar product) g. Our aim is to construct an associative algebra A with the unity 1_A, and a linear insertion $k: V \to A$ satisfying the following condition:

$$k(u)^2 = g(u, u)1_A.$$

Of the pair (A, k), we say that it is a solution of the Dirac problem. There are many such solutions, but one of them is universal in nature.

Let (e_i) be a basis of V. This basis generates a free algebra with unity, which, as a vector space, is spanned by 1 and the products of the elements of the basis $e_{i_1}, \ldots, e_{i_k}$, where $k = 1, 2, \ldots$ Next let us require that the products of the generators e_i obey the rules

$$e_i e_j + e_j e_i = 2g_{ij}.$$

In this way, we obtain the algebra $\mathbf{Cl}(V, g)$, called the *Clifford algebra* associated with the pair (V, g). If $u = u^i e_i$ and $v = v^j e_j$, by identifying these vectors with their images in the algebra $C(V, g)$, we obtain

$$u \cdot v + v \cdot u = 2g(u, v),$$

which gives in particular $u^2 = g(u, u)$. Therefore, the algebra $\mathbf{Cl}(V, g)$, together with the mapping $k(u) = u$, is a solution of the Dirac problem. Although in constructing it we used a certain basis, the construction does not in fact depend on its choice.

If the basis (e_i) is orthogonal, its different elements anticommute. Therefore, the basis of the algebra $\mathbf{Cl}(V, g)$ consists of elements of the form

$$1, e_i, e_i e_j (i < j), e_i e_j e_k (i < j < k), \ldots, e_1 \ldots e_n.$$

Thus, the dimension of $\mathbf{Cl}(V, g)$ is 2^n.

The universality of the Clifford algebra consists in the following: if (A, k) is a certain solution of the Dirac problem, there exists a homomorphism of algebras, with unity f: $\mathbf{Cl}(V, g) \to A$, such that for each vector u, $f(u) = k(u)$.

If u is not a null vector, there exists in the Clifford algebra its inverse element $u^{-1} = u/g(u, u)$. Let v be an arbitrary vector. Let us consider the expression

$$\varrho_u(v) = -uvu^{-1}.$$

Decomposing v into the parallel component $v_{||}$ and the component $v_{\perp}$ perpendicular to u,

$$v = v_{||} + v_{\perp},$$

we have

$$\varrho_u(v) = -v_{||} + v_{\perp}.$$

Therefore, $\varrho_u : V \to V$ is a reflection with respect to a subspace orthogonal to the vector u.

The Cartan theorem says that each orthogonal transformation is a composition of reflections. Thus, the general form of orthogonal transformations is as follows:

$$\varrho(u) = (-1)^k (u_1 \ldots u_k) v (u_1 \ldots u_k)^{-1},$$

where u_i satisfies $g(u_i, u_i) \neq 0$, $i = 1, \ldots, k$.

Let us now deal for a moment with the case where V is a real vector space. We can then, without loss of generality, deal only with the vectors u_i for which $g(u_i, u_i) = \pm 1$. Let $\mathbf{Pin}(V, g) \subset \mathbf{Cl}(V, g)$ be a set made up of all the elements of the form

$$s = u_1 \ldots u_k, \quad g(u_i, u_i) = \pm 1.$$

We can readily notice that the product of two such elements is of this form, while the inverse element exists and is also of this form. Thus, $\mathbf{Pin}(V, g)$ constitutes a group.

Although the factorization of s into the product of k vectors and also the number k are determined ambiguously, the parity of the element s, i.e. the number $(-1)^k = \operatorname{sgn}(s)$, is determined uniquely. In fact, the space $\mathbf{Cl}(V, g)$ decomposes into the direct sum

$$\mathbf{Cl}(V, g) = \mathbf{Cl}(V, g)^+ \oplus \mathbf{Cl}(V, g)^-$$

of the subspaces of the even elements $\mathbf{Cl}(V, g)^+$ and the odd elements $\mathbf{Cl}(V, g)^-$. Moreover, the subspace $\mathbf{Cl}(V, g)^+$ is a subalgebra of $\mathbf{Cl}(V, g)$. Therefore, the orthogonal transformation ϱ_s defined by the formula

$$\varrho_s(v) = \operatorname{sgn}(s) svs^{-1}$$

is subordinate to each element s of the group $\mathbf{Pin}(V, g)$. Furthermore, $\varrho_{s_1} \circ \varrho_{s_2} = \varrho_{s_1 s_2}$, so the mapping $s \to \varrho_s$ is a homomorphism of the group $\mathbf{Pin}(V, g)$ onto the orthogonal group $O(V, g)$.

What is the kernel of this homomorphism; i.e. when is $\varrho_s(v) = v$ for all $v \in V$? The element s belongs to the kernel if $\operatorname{sgn}(s) s e_i s^{-1} = e_i$, or

$$e_i s e_i^{-1} = \operatorname{sgn}(s) s$$

for all the elements of the orthonormal basis (e_i). We can represent the element s as a linear combination

$$s = \sum_{k=0}^{n} \sum_{i_1 < \dots < i_k} a_{i_1 \dots i_k} e_{i_1} \dots e_{i_k}$$

of the elements of the Clifford algebra. Since the vector e_i commutes with itself and anticommutes with the other elements of the basis (e_i), we have

$$e_i e_{i_1} \dots e_{i_k} e_i^{-1} = (-1)^{k-\varepsilon} e_{i_1} \dots e_{i_k}$$

where $\varepsilon = 1$ or 0, depending on whether the index i is or is not contained within the set of indices $\{i_1, \dots, i_n\}$. The sign $(-1)^{k-\varepsilon}$ does not depend on the index i only if $k = 0$ or $k = n$; therefore, the admissible form of the element s is as follows:

$$s = a + b e_1 \dots e_n, \qquad a, b \in \mathbf{R}.$$

Then

$$e_i s e_i^{-1} = a + (-1)^{n-1} b e_1 \dots e_n.$$

If n is even then $\operatorname{sgn}(s) = 1$, therefore $b = 0$. Because s must have a specified parity, there are two possibilities when n is odd: $s = a$ or $s = b e_1 \dots e_n$. In the latter case, however, $\operatorname{sgn}(s) = -1 \neq (-1)^{n-1}$. Thus, for any n, s must be a number. There are only two numbers, $+1$ and -1, within $\mathbf{Pin}(V, g)$, therefore, the kernel of the homomorphism $s \mapsto \varrho_s$ is $\{+1, -1\}$. The group $\mathbf{Pin}(V, g)$ covers the orthogonal group twice.

The group $\mathbf{Spin}(V, g)$ is defined as a subgroup of the group $\mathbf{Pin}(V, g)$ made up of its even elements. The group $\mathbf{Spin}(V, g)$ covers the subgroup $\mathbf{SO}(V, g)$ of the group $\mathbf{O}(V, g)$ twice. If the signature of the metric tensor g is strictly positive or strictly negative and if $\dim V > 1$, the group $\mathbf{Spin}(V, g)$ is connected. For the metric tensor g with arbitrary signature, we shall define its subgroup $\mathbf{Spin}_+(V, g)$ in the following way: $\mathbf{Spin}_+(V, g) = \{s \in \mathbf{Spin}(V, g) : ss^{\mathrm{T}} = 1\}$, where for $s = u_1 \dots u_k$ we define $s^{\mathrm{T}} = u_k \dots u_1$. The group $\mathbf{Spin}_+(V, g)$ covers the connected component of unity of an orthogonal group (e.g., the subgroup

$\mathbf{L}_+^\uparrow$ in the case of Minkowski space twice). If $\dim V = 1$, or if we are dealing with two-dimensional Minkowski space, this covering is trivial, i.e. $\mathbf{Spin}_+(V, g)$ decomposes into two connected components. In the opposite case, two perpendicular vector e_1 and e_2 exist in V, such that $g(e_1, e_1) = g(e_2, e_2) = \varepsilon \in \{+1, -1\}$. Then

$$s(\alpha) = \left(\varepsilon \cos\frac{\alpha}{4} e_1 + \sin\frac{\alpha}{4} e_2\right)\left(\cos\frac{\alpha}{4} e_1 - \varepsilon \sin\frac{\alpha}{4} e_2\right) = \cos\frac{\alpha}{2} + \sin\frac{\alpha}{2} e_2 e_1$$

is a continuous curve in $\mathbf{Spin}_+(V, g)$, connecting $+1 = s(0)$ with $-1 = s(2\pi)$. Therefore, $\mathbf{Spin}_+(V, g)$ is a connected group covering twice the connected component of unity of the orthogonal group. If the signature of the metric tensor g is elliptic (and $\dim V \geqslant 3$) or strictly hyperbolic (and $\dim(V \geqslant 4)$, the group $\mathbf{Spin}_+(V, g)$ is, moreover, simple connected.

The orthogonal transformation $\varrho_{s(\alpha)}$ is the rotation by the angle α in the plane spanned by the vectors e_1, e_2. Since $\varrho_{s(2\pi)} = \mathrm{id}_V$, we say that the vectors "do not feel" the sign of s. The case is different with other objects—spinors—on which the group $\mathbf{Pin}(V, g)$ acts faithfully, so that it is "only after rotation by 4π" that the spinor returns to the initial position, $s(4\pi) = 1$.

The *space of real spinors* $\Sigma_{\mathbf{R}}$ is a real vector space where a representation γ of the Clifford algebra $\mathbf{Cl}(V, g)$ acts irreducibly. To be more specific, for each $s \in C(V, g)$, $\gamma(s)$ is a linear transformation of the space $\Sigma_{\mathbf{R}}$ such that

1. $\gamma(as+bt) = a\gamma(s)+b\gamma(t)$,
2. $\gamma(st) = \gamma(s)\gamma(t)$,
3. If Σ' is a subspace of $\Sigma_{\mathbf{R}}$ and $\gamma(s)\Sigma' \subset \Sigma'$ for all $s \in \mathbf{Cl}(V, g)$ then $\Sigma' = \Sigma_{\mathbf{R}}$ or $\Sigma' = \{0\}$.

The representation γ of the algebra $\mathbf{Cl}(V, g)$ induces the representation of the group $\mathbf{Pin}(V, g)$ and its subgroups. Since non-null vectors generate both the group $\mathbf{Pin}(V, g)$| and the whole Clifford algebra, the induced representation of the group $\mathbf{Pin}(V, g)$ is irreducible.

We can now write the "square root" of the Klein–Gordon equation. Let γ be a representation of the Clifford algebra $C(V, g)$, where (V, g) is the Minkowski vector space. Denoting $\gamma_i = \gamma(e_i)$, we have

$$\gamma_i \gamma_j + \gamma_j \gamma_i = -2g_{ij}.$$

Let ψ be a real spinor field, i.e. a mapping from the Minkowski space into the space of real spinors $\Sigma_{\mathbf{R}}$. (In this case, this space is four-dimensional). Hence, the sought equation has the form

$$\left(\gamma^j \frac{\partial}{\partial x^j} - \frac{mc}{\hbar}\right)\psi = 0.$$

This equation has two essential defects. Firstly, if $m \neq 0$, there is no Lagrangian formulation of this equation which would be invariant with respect to the group $\mathbf{Spin}_+(V, g)$. Secondly, the interaction of each particle field with the electromagnetic field is included in the equations of this field by replacing $\frac{\partial}{\partial x^j}$ with the expression $\frac{\partial}{\partial x^j} - \mathrm{i}\frac{e}{\hbar c} A_j$, where A_j is the four-potential of the electromagnetic field. To do this, we require the complexification of the Clifford algebra and the spinor space. The complexifield Clifford algebra is equal to the Clifford algebra of the complexifield space $V^{\mathbf{C}}$, with the scalar product g in $V^{\mathbf{C}}$ determined according to the prescription in Chapter 7. Taking ψ as a field with values in the space of complex spinors, we call the above equation the Dirac equation. In the complex case, an invariant Lagrangian exists.

If the dimension n of the space V is even, $n = 2\nu$, the algebra $\mathbf{Cl}(V, g)^{\mathbf{C}}$ is isomorphic to the algebra $\mathbf{C}(2^\nu)$ of all complex square matrices of degree 2^ν. All the irreducible representations of this algebra are mutually equivalent, while the dimension of the space of the representation (i.e. the space of complex spinors) is 2^ν. Let (e_i) be a basis in $V^{\mathbf{C}}$, such that $g(e_i, e_j) = \delta_{ij}$, and let

$$\gamma_{n+1} = i^\nu \gamma(e_1) \ldots \gamma(e_n).$$

Then γ_{n+1} has the following properties: 1. γ_{n+1} anticommutes with each vector $\gamma(u)$, $\gamma_{n+1}\gamma(u)+\gamma(u)\gamma_{n+1} = 0$, 2. $\gamma_{n+1}^2 = 1$. We shall define two subspaces $\Sigma_\pm = \{\xi \in \Sigma: \gamma_{n+1}\xi = \pm\xi\}$ of the space of spinors Σ, i.e. the spaces of left and right half-spinors. We can represent each spinor as the sum

$$\xi = \xi_+ + \xi_-,$$

where $\xi_\pm = \frac{1 \pm \gamma_{n+1}}{2}\xi$ and, as we can easily check, $\xi_\pm \in \Sigma_\pm$. For a non-null vector u, $\gamma_{n+1} = -\gamma(u)\gamma_{n+1}\gamma(u)^{-1}$, and hence $\mathrm{Tr}\,\gamma_{n+1} = -\mathrm{Tr}\,\gamma_{n+1}$, i.e. $\mathrm{Tr}\,\gamma_{n+1} = 0$. (Similarly, we can demonstrate that for each orthogonal basis (e_i) the trace of the linear transformations $\gamma(e_i)$ and their products which are not numbers is 0). Thus, the dimensions of the spaces Σ_+ and Σ_- must be equal to each other, $\dim\Sigma_+ = \dim\Sigma_- = 2^{\nu-1}$. In the spaces of half-spinors Σ_+ and Σ_-, the representation γ induces irreducible and mutually nonequivalent representations of the algebra $\mathbf{Cl}^+(V, g)$ and of the groups $\mathbf{Spin}(V, g)$ and $\mathbf{Spin}_+(V, g)$.

If the dimension n of the space V is odd, $n = 2\nu+1$, we can form two irreducible representations of the algebra $\mathbf{Cl}(V^{\mathbf{C}}, g)$ in the following way. Let (e_i) be a certain orthonormal basis of the space $V^{\mathbf{C}}$, $g(e_i, e_j) = \delta_{ij}$, and let V' be a subspace of the space $V^{\mathbf{C}}$ spanned by the vectors $e_1, \ldots, e_{2\nu}$. Let us construct the Clifford algebra of this subspace and the representation γ of this algebra in the spinor space $\Sigma \simeq \mathbf{C}^{2^\nu}$. In Σ, we define the two representations γ_+, γ_- of the algebra $\mathbf{Cl}(V^{\mathbf{C}}, g)$ by the formulae

$$\gamma_+(e_i) = \gamma(e_i), \quad i = 1, \ldots, 2\nu,$$

$$\gamma_+(e_{2\nu+1}) = i^\nu \gamma(e_i) \dots \gamma(e_{2\nu}),$$

and

$$\gamma_-(e_i) = \gamma(e_i), \qquad i = 1, \dots, 2\nu,$$

$$\gamma_-[e_{2\nu+1}) = -i^\nu \gamma(e_1) \dots \gamma(e_{2\nu}).$$

These representations are irreducible, unfaithful and mutually nonequivalent. Each irreducible representation of the algebra $\mathbf{Cl}(V^{\mathbf{C}}, g)$ is equivalent to one of the representations γ_+ and γ_-. In the space $\Sigma \oplus \Sigma$ we can create a faithful representation Γ of the algebra $\mathbf{Cl}(V^{\mathbf{C}}, g)$ in the block-diagonal form

$$\Gamma(s) = \begin{pmatrix} \gamma_+(s) & 0 \\ 0 & \gamma_-(s) \end{pmatrix}, \qquad s \in \mathbf{Cl}(V^{\mathbf{C}}, g).$$

In contrast to the even-dimensional case, all the irreducible representations of the algebra $\mathbf{Cl}(V^{\mathbf{C}}, g)$ and of the groups $\mathbf{Spin}(V, g)$ and $\mathbf{Spin}_+(V, g)$ are equivalent. In particular, this is the case with the representations induced by γ_+ and γ_-.

Let us now concentrate on the Minkowski space (V, g). We shall give the matrix representation of the Clifford algebra $\mathbf{Cl}(V, g)^{\mathbf{C}}$. Let

$$\varepsilon = \begin{pmatrix} 0 & 1 \\ -1 & 0 \end{pmatrix}.$$

This matrix has the following properties: 1. for each matrix A, $A^{\mathrm{T}} \varepsilon A = \det A \cdot \varepsilon$, 2. $\varepsilon^2 = -1$, 3. $\det \varepsilon = 1$. We define $\gamma: V^{\mathbf{C}} \to \mathbf{C}(4)$ by the following formula

$$\gamma(u) = \begin{pmatrix} 0 & \sigma(u)^{\mathrm{T}} \\ -\sigma(u) & 0 \end{pmatrix}.$$

We can easily check that $\gamma(u)^2 = \det \sigma(u) = g(u, u)$, as it should. In this way, we have constructed the representation $\mathbf{Cl}(V^{\mathbf{C}}, g)$ in $\Sigma = \mathbf{C}^4$.

The element s of the group $\mathbf{Spin}(V, g)$

$$s = u_1 \dots u_{2k}$$

has the representation

$$\gamma(s) = \begin{pmatrix} U(s) & 0 \\ 0 & V(s) \end{pmatrix},$$

where

$$U(s) = (-1)^k \sigma(u_1)^{\mathrm{T}} \varepsilon \sigma(u_2) \varepsilon \dots \sigma(u_{2k-1})^{\mathrm{T}} \varepsilon \sigma(u_{2k}) \varepsilon,$$

$$V(s) = (-1)^k \sigma(u_1) \varepsilon \sigma(u_2)^{\mathrm{T}} \dots \sigma(u_{2k-1}) \varepsilon \sigma(u_{2k})^{\mathrm{T}} \varepsilon.$$

Since for real vectors $\sigma(u)^{\mathrm{T}} = \overline{\sigma(u)}$, we find that $V(s) = \overline{U(s)}$ and $\det U(s) = s^{\mathrm{T}} s = \pm 1$. The action of the group $\mathbf{Spin}(V, g)$ on the vectors is given by the formula

$$\gamma(s)\gamma(u)\gamma(s^{-1}) = \begin{pmatrix} 0 & \det U \cdot (\bar{U}\sigma(u) U^{\mathrm{T}}\varepsilon \\ -\det U \cdot \bar{U}\sigma(u) U^{\mathrm{T}}\varepsilon & 0 \end{pmatrix}.$$

and thus

$$\sigma(u) \to \det U \cdot \bar{U}\sigma(u) U^{\mathrm{T}}.$$

If $s \in \mathbf{Spin}_+(V, g)$, this action is in agreement with that of the group $\mathbf{SL}(2, \mathbf{C})$ discussed in Chapter 7.

To cover the group $\mathbf{L}_+^\uparrow$, the group $\gamma(\mathbf{Spin}_+(V, g))$ must contain all the matrices

$$\begin{pmatrix} U & 0 \\ 0 & \bar{U} \end{pmatrix},$$

such that $\det U = 1$. We have thus shown that there is an isomorphism between the groups $\mathbf{Spin}_+(V, g)$ and $\mathbf{SL}(2, \mathbf{C})$.

The spinor space is the direct sum of the spaces Σ_+ and Σ_-. In the present representation, we can regard the elements of Σ_+ as spinors with the last two components equal to 0, and the elements of Σ_- as spinors with the first two components different from 0. Let us denote the components of the half-spinor $\varphi \in \Sigma_+$ as ψ^A (where $A, B, \ldots = 1, 2$), and the components of the half-spinor $\varphi \in \Sigma_-$ as $\varphi^{\dot{A}}(\dot{A}, \dot{B}, \ldots = 1, 2)$. The group $\mathbf{Spin}_+(V, g)$ acts in Σ_+ in the following way:

$$\psi^A \mapsto U^A_B \psi^B,$$

and in the space Σ_- in the following way:

$$\varphi^{\dot{A}} \mapsto \bar{U}^{\dot{A}}_{\dot{B}} \varphi^{\dot{B}},$$

where $\bar{U}^{\dot{A}}_{\dot{B}} = \overline{U^A_B}$. Further on, we shall consider that complex conjugation transfers from one space of half-spinors to the other and denotes $\bar{\psi}^{\dot{A}} = \overline{\varphi^A}, \bar{\psi}^A = \overline{\varphi^{\dot{A}}}$.

The matrices $U = (U^A_B) \in \mathbf{SL}(2, \mathbf{C})$ have the following property:

$$U^A_C \varepsilon_{AB} U^B_D = \varepsilon_{CD},$$

where $(\varepsilon_{AB}) = \varepsilon$. It follows from this property that the bilinear form in the space Σ_+

$$\varepsilon_{AB} \varphi^A \psi^B$$

is invariant with respect to the action of the group $\mathbf{Spin}_+(V, g)$. Its complex conjugate, which is a bilinear form in the space Σ_-,

$$\varepsilon_{\dot{A}\dot{B}} \varphi^{\dot{A}} \varphi^{\dot{B}},$$

is also invariant. $\varepsilon_{\dot{A}\dot{B}}$ is numerically equal to ε_{AB}, ε^{AB} and $\varepsilon^{\dot{A}\dot{B}}$ also denote components of the matrix ε.

From the spaces $\Sigma_+$and Σ_- we can form their tensor products $\Sigma^{k,l}$. An element of such a product has the form $\psi^{A_1 \dots A_k \dot{B}_1 \dots \dot{B}_l}$, and the group $\mathbf{Spin}_+(V, g)$ acts on it the following way:

$$\psi^{A_1 \dots A_k \dot{B}_1 \dots \dot{B}_l} \mapsto U^{A_1}_{C_1} \dots U^{A_k}_{C_k} \overline{U}^{\dot{B}_1}_{\dot{D}_1} \dots \overline{U}^{\dot{B}_l}_{\dot{D}_l} \psi^{C_1 \dots C_k \dot{D}_1 \dots \dot{D}_l}$$

where this action commutes with the complex conjugation transferring $\Sigma^{l,k}$ onto $\Sigma^{k,l}$. If $k = l$, the complex conjugation leaves spinors in the same space $\Sigma^{k,k}$. If the element $\psi \in \Sigma^{k,k}$ does not change, we call it Hermitian.

An interesting part of the space $\Sigma^{k,l}$ is the subspace $S^{k,l}$ composed of spinors symmetric in both kinds of indices, i.e. its elements satisfy the equalities

$$\psi^{A_1 \dots A_k \dot{B}_1 \dots \dot{B}_l} = \psi^{(A_1 \dots A_k) \dot{B}_1 \dots \dot{B}_l} = \psi^{A_1 \dots A_k (\dot{B}_1 \dots \dot{B}_l)}.$$

In these spaces the group $\mathbf{Spin}_+(V, g)$ acts in an irreducible way, i.e. the spaces $S^{k,l}$ do not contain nontrivial subspaces which are invariant under the action of this group. It turns out that each finite-dimensional, irreducible representation of this group is equivalent to one of these representations. If $k+l$ is an even number, the representation does not distinguish the sign of the element $s \in \mathbf{Spin}_+(V, g)$. It is then and only then that the elements of the space $S^{k,l}$ have tensor counterparts.

It follows from the form of $\gamma(u)$ that $\sigma(u)\varepsilon$ is a linear operation from the space of half-spinors $\Sigma_- = \Sigma^{0,1} = S^{1,0}$ into the space $\Sigma_+ = \Sigma^{1,0} = S^{1,0}$, and, more over, this operation commutes with the action of the group $\mathbf{Spin}_+(V, g)$ in these spaces. If $\varphi = (\varphi^A) \in \Sigma_-$, then

$$(\sigma(u)\varepsilon\varphi)^A = \sigma(u)^{A\dot{B}} \varepsilon_{\dot{B}\dot{C}} \varphi^{\dot{C}}.$$

Therefore, the element

$$\sigma(u)^{A\dot{B}} = \sigma^{A\dot{B}}_k u^k,$$

where $(\sigma^{A\dot{B}}_k)$ are Pauli matrices ($k = 0, 1, 2, 3$), belongs to the space $\Sigma^{1,1} = S^{1,1}$. Moreover, it is Hermitian if and only if the vector u is real. Below, we continue to denote it as $u^{A\dot{B}} = \sigma(u)^{A\dot{B}}$ and call it the *spinor image of the vector u*. The inverse formula has the form

$$u^k = \sigma^k_{A\dot{B}} u^{A\dot{B}}, \qquad \text{where} \qquad \sigma^k_{A\dot{B}} = g^{kj} \varepsilon_{AC} \varepsilon_{\dot{B}\dot{D}} \sigma^{C\dot{D}}_j.$$

What does the spinor image of a null vector look like? In Chapter 7, we said that $\det \sigma(u) = 0$ if and only if the matrix $\sigma(u)$ factorizes, i.e. $u^{A\dot{B}}$ has the form

$$u^{A\dot{B}} = \alpha^A \beta^{\dot{B}}.$$

In the particular case where the vector u is real, to ensure that $u^{A\dot{B}}$ is Hermitian the spinors β^A and α^A must be proportional, $\beta^A = \lambda\alpha^A$, and, moreover, λ should be real. Rescaling the spinor α^A, we can thus always obtain the following representation of the real null vector

$$u^{A\dot{B}} = \pm \alpha^A \overline{\alpha}^{\dot{B}},$$

and, furthermore, we can find that the $+$sign corresponds to vectors directed to the future, and the $-$sign is for vectors directed to the past (assuming here that the vector e_0 was directed to the future).

Just as the spinor image of a vector, we can create the spinor image of any tensor. Let us consider, e.g., the spinor image of an antisymmetric tensor of second order, $F^{ij} = -F^{ji}$. It is given by the formula

$$F^{AB\dot{C}\dot{D}} = \sigma_i^{A\dot{C}}\sigma_j^{B\dot{D}}F^{ij}.$$

The fact that F^{ij} is antisymmetric gives $F^{AB\dot{C}\dot{D}} = -F^{BA\dot{D}\dot{C}}$, and, hence, the following decomposition is valid

$$F^{AB\dot{C}\dot{D}} = \frac{1}{2}(F^{AB\dot{C}\dot{D}} - F^{AB\dot{D}\dot{C}}) + \frac{1}{2}(F^{AB\dot{C}\dot{D}} - F^{BA\dot{C}\dot{D}}).$$

The first term of this decomposition is antisymmetric in the indices $\dot{C}\dot{D}$. Since in a two-dimensional space each object which is antisymmetric in two indices is proportional to $\varepsilon^{\dot{C}\dot{D}}$, this means that

$$\frac{1}{2}(F^{AB\dot{C}\dot{D}} - F^{AB\dot{D}\dot{C}}) = F^{AB}\varepsilon^{\dot{C}\dot{D}}.$$

The spinor F^{AB} defined by this equation is, as follows from the condition $F^{AB\dot{C}\dot{D}} = -F^{BA\dot{D}\dot{C}}$, symmetric: $F^{AB} = F^{BA}$. Similarly, we can write the second term of the decomposition as

$$\frac{1}{2}(F^{AB\dot{C}\dot{D}} - F^{BA\dot{C}\dot{D}}) = G^{\dot{C}\dot{D}}\varepsilon^{AB},$$

where $G^{\dot{C}\dot{D}}$ is symmetric. Thus,

$$F^{AB\dot{C}\dot{D}} = F^{AB}\varepsilon^{\dot{C}\dot{D}} + G^{\dot{C}\dot{D}}\varepsilon^{AB}.$$

Conversely, a certain antisymmetric complex tensor F^{ij} is connected with each pair of symmetric spinors $(F^{AB}, G^{\dot{C}\dot{D}})$. A special situation arises if the tensor F^{ij} is real. Then, in order that $F^{AB\dot{C}\dot{D}}$ may be Hermitian, $G^{\dot{A}\dot{B}} = \overline{F^{AB}} = \bar{F}^{\dot{A}\dot{B}}$ must hold. Thus, the spinor image of an electromagnetic field tensor is the symmetric spinor F^{AB}. It is interesting to note the form of Maxwell's vacuum equations in this formalism:

$$\frac{\partial}{\partial x^{A\dot{B}}}F^{AC} = 0.$$

We may now reverse the questions which we have posed so far and ask if we can give the vector, or tensor, image of a half-spinor. Since the group $\mathbf{Spin}_+(V, g)$, acting on half-spinors, distinguishes the signs of the elements of this group, and fails to do so on vectors and tensors, there is no mutually unique tensor image of a spinor. An image exists however, in which the same tensor counterpart is attributed to the half-spinors α^A and $-\alpha^A$ (and only to them). W can define

$$k^{A\dot{B}} = \alpha^A\bar{\alpha}^{\dot{B}}, \qquad F^{AB} = \alpha^A\alpha^B,$$

and then the pair (k^i, F^{ij}) defines the spinor α^A to within the sign. The tensor F^{ij} is given by the formula

$$F^{ij} = \sigma^i_{A\dot{B}}\sigma^j_{C\dot{D}}(\varepsilon^{AC}\alpha^{\dot{B}}\alpha^{\dot{D}} + \varepsilon^{\dot{B}\dot{D}}\alpha^A\alpha^C).$$

Let us define the auxiliary half-spinor β^A by the formula

$$\varepsilon^{AC} = \alpha^A\beta^C - \beta^A\alpha^C.$$

It is determined to within a transformation $\beta^A \mapsto \beta^A + \lambda\alpha^A$. Substituting ε^{AC} in the formula for F^{ij}, we obtain, after simple transformations,

$$F^{ij} = k^ie^j - e^ik^j,$$

where

$$e^i = \sigma^i_{A\dot{B}}(\beta^A\alpha^{\dot{B}} + \beta^{\dot{B}}\alpha^A)$$

is a real vector orthogonal to the vector k^i. It is given to within a transformation $e^i \mapsto e^i + (\lambda + \bar{\lambda})k^i$. The tensor F^{ij} in the given form corresponds to the electromagnetic field of a plane wave; the null vector k^i is proportional to the wave four-vector of this wave.

Let us now consider spinors of the $S^{k,0}$ type. It turns out that the fields of particles with 0 mass and the spin $k\hbar/2$ are fields with values in these spaces. The free equations of these fields equations (Pauli–Fierz) have the form

$$\frac{\partial}{\partial x^{A_1\dot{B}}} F^{A_1\dots A_k} = 0;$$

they are, therefore, certain generalizations of Maxwell's equations.

Let us consider the function

$$F(z_A) = F^{A_1\dots A_k}z_{A_1}\dots z_{A_k},$$

which is a homogeneous polynomial of the kth degree of the varables z_1 and z_2. It follows from the fundamental theorem of algebra that we can factorize this function,

$$F^{A_1\dots A_k}z_{A_1}\dots z_{A_k} = (\alpha^{A_1}z_{A_1})\dots(\varkappa^{A_k}z_{A_k}),$$

which implies in turn that

$$F^{A_1\dots A_k} = \alpha^{(A_1}\dots\varkappa^{A_k)}.$$

This decomposition of the spinor $F^{A_1\dots A_k}$ is called *canonical.* The spinors $\alpha^A, \dots, \varkappa^A$, which occur in the above formula are defined to within a coefficient, and only their directions are determined uniquely. Since a certain null vector is connected with each half-spinor, the spinor $F^{A_1\dots A_k}$ determines k null directions in Minkowski space, called its *principal directions.*

The canonical decomposition of the spinor of electromagnetic field is related to the algebraic classification of this field; namely, we say that F^{ij} is of the general type, if the principal directions of the spinor F^{AB} are different, i.e.

$$F^{AB} = \alpha^{(A}\beta^{B)}, \qquad \beta^A \neq \alpha^A$$

and that it is of the null type (or the plane-wave type), if these directions coincide, i.e.

$$F^{AB} = \alpha^A\alpha^B.$$

Of course, this classification is connected with a single point in spacetime and can vary from point to point.

CHAPTER 10

Newtonian Theory of Gravitation and the Principle of Equivalence

We now pass on to a discussion of spacetime models which take into account gravitational phenomena. Why is it that gravity plays so exceptional a role as to be specially considered in developing a spacetime model, while the electromagnetic or nuclear phenomena "superpose" themselves, as it were, on that model? We have already tried to answer this question in Chapter 3. Now we shall examine it in more detail. We will begin at the Newtonian level, without reference to the results of special relativity.

It is known that the mass of a body appears in at least three different roles, namely as

(1) *inertial mass*, m_I,

(2) *active gravitational mass*, m_{AG},

(3) *passive gravitational mass*, m_{PG}.

The inertial mass occurs in Newton's equation $m_I \ddot{\mathbf{r}} = \mathbf{F}$.

The active gravitational mass is the source of a gravitational field and appears in Poisson's equation for the gravitational potential φ,

$$\Delta\varphi = 4\pi k m_{AG}\,\delta(\mathbf{r}-\mathbf{r}_1(t)),$$

where the function $\mathbf{r}_1(t)$ describes the motion of the body, and k is the gravitational constant. This equation has a solution

$$\varphi = -\frac{km_{AG}}{r},$$

where $r = |\mathbf{r}-\mathbf{r}_1|$.

The passive gravitational mass occurs in the expression for the force

$$\mathbf{F} = -m_{PG}\,\mathrm{grad}\,\varphi.$$

The third law of dynamics ensures the equality of the two gravitational masses. Indeed, it implies that

$$\overset{1}{m}_{AG}\overset{2}{m}_{PG} = \overset{1}{m}_{PG}\overset{2}{m}_{AG},$$

which can be written

$$\frac{\overset{1}{m}_{AG}}{\overset{1}{m}_{PG}} = \frac{\overset{2}{m}_{AG}}{\overset{2}{m}_{PG}} = \lambda = \text{const}.$$

Negation of the above assertion would lead to the conclusion that the total momentum of the system consisting of bodies 1 and 2 is not conserved. That the constant λ may be different from 1 is unessential since it can be absorbed into the gravitational constant. Thus, we have

$$m_{PG} = m_{PG} = m_G.$$

The gravitational mass m_G plays a similar role to that of an electric charge in electrodynamics, and is often spoken of as gravitational charge. An essential difference between gravitation and electrodynamics lies in the equality $m_I = m_G$. This equality is an experimental fact, which has been verified to a very high degree of accuracy. In 1894 Eötvös achieved an accuracy of the order of 10^{-8} (although it is now believed that his measurements were in fact somewhat less accurate). In 1964 Dicke and his collaborators reached the accuracy of 10^{-11}, which was improved to 10^{-12} by Braginski in 1971.

Accepting the equality $m_I = m_G$ as exact leads to the declaration that the equation of motion of a body in a gravitational field is universal. This means that all bodies in that field satisfy the relationship

$$\ddot{\mathbf{r}} + \operatorname{grad} \varphi = 0.$$

It follows that all bodies starting with the same initial conditions move in the gravitational field in the same way. The formulation of the first law of dynamics must be changed if we want to take gravity into account. We will now say that there exists a class of preferred motions, called free falls, a reference frame, and a function φ, such that the equation of free falls has the form written above.

What was underlying the affine structure of gravity-free spacetime? Of course it was the fact that the equation of motion of free particles was of the form

$$\ddot{\mathbf{r}} = 0,$$

which is equivalent to the equation of a straight line

$$\mathbf{r} = \mathbf{r}_0 + \mathbf{V}t.$$

The affine structure of spacetime ties in closely with the geometric representation of free motions. Now, with the changed equation of free particle motion, it may be doubted whether spacetime can be realized as an affine space. For instance, we can no longer assume that free falls correspond to straight lines in E, because examples can readily be given of falls whose world-

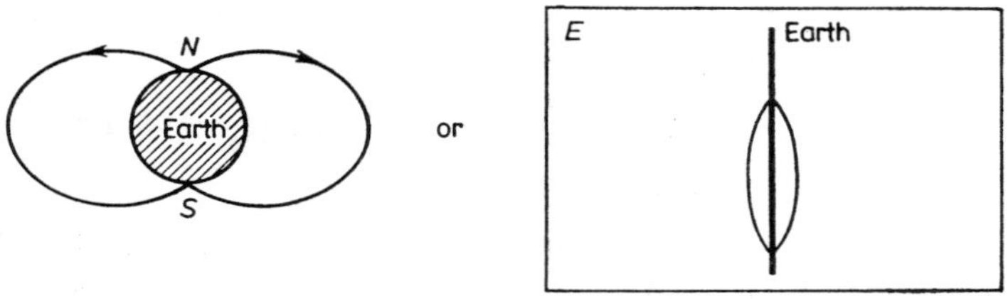

Fig. 10.1

lines intersect at more than one point (Fig. 10.1). (Here the possibility of identifying the world-lines of free falls with the extremal lines of the corresponding curved geometry suggests itself). It may appear that a way to save the affine structure is to accept that free falls are represented by curves, while straight lines correspond to some other, fictitious motions:

$$\ddot{\mathbf{r}}+\operatorname{grad}\varphi = 0 \leftrightarrow \text{certain curves in } E,$$

$$\mathbf{r} = \mathbf{r}_0+\mathbf{V}t \quad \leftrightarrow \text{straight lines in } E.$$

It is easy to see, however, that the form of the equation of free falls remains unchanged after transformation

$$\mathbf{r} \mapsto \mathbf{r} + \mathbf{a}(t)$$

$$\varphi \mapsto \varphi - \ddot{\mathbf{a}}\cdot\mathbf{r},$$

where $\mathbf{a}(t)$ is an arbitrary vector function of time. Under this transformation, owing to the arbitrariness of $\mathbf{a}(t)$, the linear relations in the equation of a straight line will become nonlinear. The affine structure defined above would not be independent of the choice of reference system.

The formulae for the transformation of position and potential reflect our inability to distinguish between gravitational and inertial forces by means of local experiments. This is illustrated by "Einstein's elevator" (Fig. 10.2).

An observer in an elevator which is at rest (in the Earth's gravitational field) and an observer in a rocket that moves with a uniform acceleration are in similar situations. The enclosed observer cannot tell in which of these situations he actually is. The equivalence of gravitational and inertial forces becomes even more apparent when we consider an elevator (or rocket) accelerating in the gravitation field of the Earth. An observer conducting local experiments is not able to separate gravitation from inertia.

The requirement of localization is essential. To reject it, one could demand that the potential φ vanish far away from the source of the gravitational field, i.e. $\varphi \to 0$ as $r \to \infty$. Of course this cannot be done in a closed elevator, because it would require the introduction of a global reference frame and the observation of events at large distances from the bodies creating the field.

Neither can the condition of the vanishing of φ at infinity be satisfied in

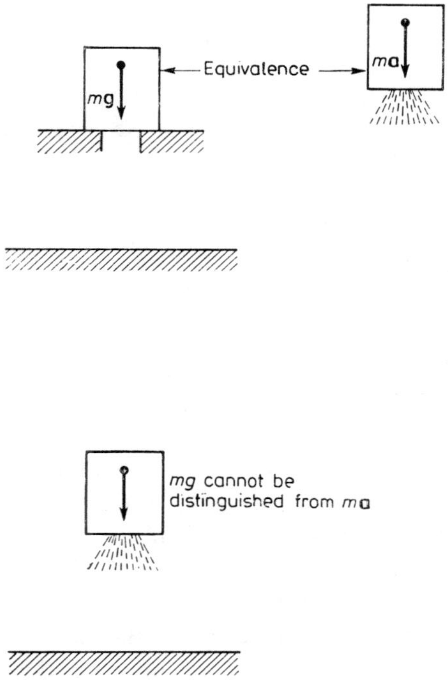

Fig. 10.2

cosmology, where one has to allow for the approximately homogeneous distribution of matter in the Universe.

Assuming that the mass density is constant throughout the Universe, we obtain for the mean gravitational potential the equation

$$\Delta\varphi = \text{const},$$

whose solution is $\varphi \sim r^2$. Thus no preferred inertial system can be distinguished in this case either. In Newtonian cosmology one should accept the complete equivalence of all frames in which the equation of free falls holds, $\ddot{\mathbf{r}} + \operatorname{grad}\varphi = 0$.

The final conclusion from these considerations is that when gravity is taken into account, it is impossible to introduce—by invoking the laws of mechanics—an affine structure into spacetime. In other words, one cannot introduce inertial systems. As in the case of the principle of relativity, we shall postulate after Einstein that this cannot be done by any experiments, not only mechanical ones. This generalization is not trivial, considering the absence of relevant experimental data.

The situation can be summarized in a table:

	Newton's physics	Einstein's physics
Without gravity	No preferred inertial frames can be distinguished by mechanical experiments	No preferred inertial frames can be distinguished by any experiment
With gravity	Inertial frames cannot be introduced by local mechanical experiments	Inertial frames cannot be introduced by any local experiment

We recall the importance of the word "local" here; with it, the theory of relativity does not satisfy Mach's principle, which asserts, in essence, that there exists a close relationship between local laws and phenomena and the motion and distribution of matter in the Universe as a whole.

How can this negative postulate be used as a basis for constructing a theory of gravity? Let us think of the way from Galilean theory to special relativity, which was also based on negative assertion. In Galilean spacetime (or Maxwell–Lorentz's) we had at our disposal the absolute time form τ, the scalar product h and the ether e (or the ether e and the scalar product g). The step to special relativity consisted in giving up the ether. Schematically this step may be represented as

$$(h, \tau, e) \leftrightarrow (e, g) \to g.$$

Now we shall analyse a different aspect of spacetime's structure. We shall see that every affine space admits an object Γ called the affine connection, and that certain restrictions on this connection (flatness) imply the local affinity of spacetime. Since, as we have already seen, spacetime cannot be affine if gravity is allowed, it will be appropriate not to assume the flatness of the affine connection. This situation can be described by a diagram analogous to the previous one:

$$(E, V) \leftrightarrow (E, \text{flat } \Gamma) \to (E, \Gamma).$$

Let us note, in passing, that the assertion of local indistinguishability of the

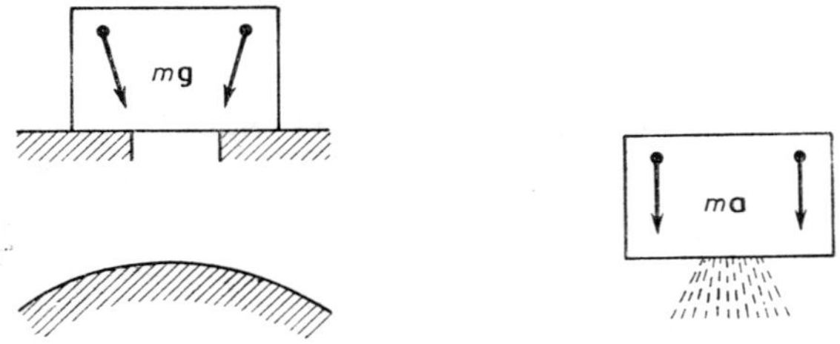

Fig. 103.

forces of inertia from the forces of gravity has limited application. The paths of two particles falling in an elevator cabin (Fig. 10.3) will approach each other because the particles move towards the centre of the Earth; this effect will not be observed in the case of particles in an accelerating rocket. Under the transformation

$$\varphi \mapsto \varphi - \ddot{\mathbf{a}} \cdot \mathbf{r}$$

we have

$$\operatorname{grad} \varphi \mapsto \operatorname{grad} \varphi - \ddot{\mathbf{a}},$$

but the second derivatives of the potential remain unchanged:

$$\frac{\partial^2 \varphi}{\partial x^\alpha \partial x^\beta} \mapsto \frac{\partial^2 \varphi}{\partial x^\alpha \partial x^\beta}.$$

It is the second derivatives that reflect the particles' drawing closer, and it is known that the second derivatives of the potential are different from zero only in real gravitational fields.

For let us consider a certain free fall $x^\alpha(t)$ and the family of the adjacent free falls $x^\alpha(t, \eta)$, such that $x^\alpha(t, \eta = 0) = x^\alpha(t)$. We call the vector $n^\alpha = \left.\frac{\partial x^\alpha}{\partial \eta}\right|_{\eta=0}$ the separation vector of this family. With accuracy up to the first-order terms in η, the vector $\eta n^\alpha(t)$ connects the points of adjacent falls at a time t: the standard fall $x^\alpha(t)$ and the fall $x^\alpha(t, \eta)$.

To find an equation which satisfies the separation vector, we differentiate with respect to η at $\eta = 0$ the equation of free falls for the whole family:

$$\frac{\partial^2 x^\alpha}{\partial t^2} + \frac{\partial \varphi}{\partial x^\alpha} = 0.$$

Substituting the partial derivatives and observing that

$$\left.\frac{\partial}{\partial \eta}\right|_{\eta=0} = n^\alpha \frac{\partial}{\partial x^\alpha},$$

we obtain the formula

$$\frac{\mathrm{d}^2 n^\alpha}{\mathrm{d}t^2} + \frac{\partial^2 \varphi}{\partial x^\alpha \partial x^\beta} n^\beta = 0.$$

Thus the second derivatives of the potential determine the relative acceleration of free falling particles. In apparent gravitational fields, the second derivatives of the potential vanish, causing the fact that the relative velocity of adjacent particles remains the same. In the next chapter we shall consider the counterpart of this equation in the general relativity theory.

CHAPTER 11

Geometric Foundations of the General Theory of Relativity

Recall that the equality of gravitational and inertial masses, $m_G = m_I$, implies the impossibility of defining inertial reference systems by means of mechanical experiments and, according to Einstein; this impossibility holds even if non-mechanical experiments are allowed. On the other hand, we know that the fundamental laws of physics can be expressed in terms of differential equations; this property is referred to as the "infinitesimal locality" of the physical laws. This implies, in turn, that spacetime should be "infinitesimally affine". In other words, the new model of spacetime should preserve of the structure of an affine space those elements that allow us to evaluate derivatives of geometric objects used to describe physical quantities.

We shall now sketch the concept of an affine connection in an n-dimensional affine space (E, V).

A frame $(\mathfrak{o}, e_i)$ defines rectilinear coordinates $\xi: E \to \mathbf{R}^n$ by means of the equation

$$p = \xi^i(p)e_i + \mathfrak{o}.$$

They are called rectilinear coordinates, since fixing $(n-1)$ from among them determines a straight line in E.

For the curve $q: \mathbf{R} \to R$ we can write

$$q(\lambda) = q^i(\lambda)e_i + \mathfrak{o}.$$

The vector tangent to the curve q is given by

$$u(\lambda) = \lim_{\Delta\lambda \to 0} \frac{q(\lambda+\Delta\lambda)-q(\lambda)}{\Delta\lambda} = \frac{dq^i}{d\lambda} e_i.$$

For an affinely parametrized straight line, the tangent vector $u(\lambda)$ is constant.

Let us now introduce curvilinear coordinates in E. Let the mapping $f: \mathbf{R}^n \to \mathbf{R}^n$ be a diffeomorphism, i.e. a bijection such that f and f^{-1} are continuously differentiable (any number of times). Let us determine the mapping $\bar{\xi} = f \circ \xi: E \to \mathbf{R}^n$. We can see that the pair $(E, \bar{\xi})$ is a chart in E; $\bar{\xi}$ is called the *curvilinear coordinate system in E*. Let us consider i-th line of the coordinate system $\bar{\xi}$, i.e., the curve q_i the image of which in $\mathbf{R}^n$ is

$$\bar{\xi}^j(q_i(\lambda)) = \begin{cases} \bar{\xi}_0^j = \text{const} & \text{for } j \neq i, \\ \lambda & \text{for } j = i. \end{cases}$$

At this point, we must make a remark concerning the notation used in physics. The mapping

$$f: \mathbf{R}^n \to \mathbf{R}^n$$

is usually written as

$$\bar{x} = f(x) \quad \text{or} \quad \bar{x} = \bar{x}(x),$$

while the inverse mapping f^{-1} is

$$x = f^{-1}(\bar{x}) \quad \text{or} \quad x = x(\bar{x}).$$

We have

$$q_i(\lambda) = \xi^j(q_i(\lambda))e_j + \mathfrak{o},$$

but $\xi = f^{-1} \circ \bar{\xi} = x \circ \bar{\xi}$, so

$$q_i(\lambda) = x^j \circ \bar{\xi}(q_i(\lambda))e_j + \mathfrak{o}$$
$$= x^j(\bar{\xi}_0^1, \ldots, \bar{\xi}_0^{i-1}, \lambda, \bar{\xi}_0^{i+1}, \ldots, \bar{\xi}_0^n)e_j + \mathfrak{o},$$

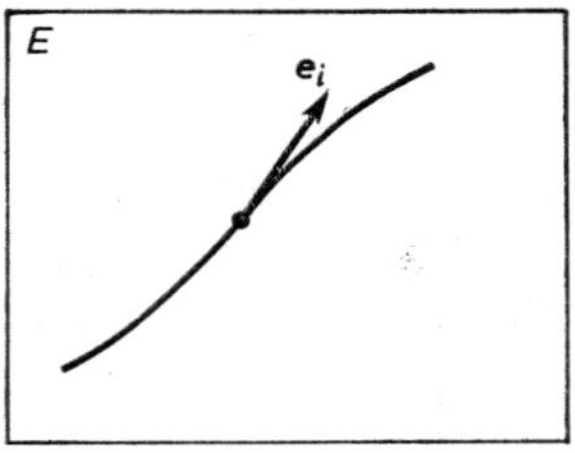

Fig. 11.1

therefore, the tangent vector to the line q_i (Fig. 11.1) is equal to

$$\bar{e}_i = \frac{dq_i}{d\lambda} = \frac{\partial x^j}{\partial \bar{x}^i} e_j,$$

where $\bar{e}_i$ is a vector field defined on E. Since f is by assumption a diffeomorphism $\det(\partial x^j/\partial \bar{x}^i) \neq 0$, and, therefore $(\bar{e}_i(p))$ is a basis in V determined by the chart $(E, \bar{\xi})$ (for any point $p \in E$).

We can decompose any vector $v \in V$ with respect to this basis:

$$v = \bar{v}^i \bar{e}_i,$$

(where, if v is the vector field $v: E \to V$, it is reasonable to decompose the vector $v(p)$ with respect to a basis at point $p \in E$). From the fact $v = v^i e_i$ and from the relation between $\bar{e}_i$ and e_i, the following equation follows:

$$\bar{v}^i = \frac{\partial \bar{x}^i}{\partial x^j} v^j.$$

We can see that a transition from one chart to another (not necessarily from a rectilinear chart to a curvilinear one) corresponds to a change in the basis at all points of the manifold E. This change in turn leads to a transformation of the vector components in agreement with the formula given above. It turns out that the components of other geometric objects are transformed in a similar way.

Let (e^i) be a basis dual to the basis (e_i), i.e. a basis in the space of linear forms (covectors) V^*, the action of which on the elements of the basis (e_i) may be expressed by the formula

$$\langle e_i, e^j \rangle = \delta_i^j .$$

The basis $(\bar{e}^i)$ dual to the transformed basis $(\bar{e}_i)$ is related, as we can readily find out, to the basis (e^i) by the formula

$$\bar{e}^i = \frac{\partial \bar{x}^i}{\partial x^j} e^j .$$

Hence, in a similar way as that used for vectors, we obtain the transformation law for covectors:

$$\bar{w}_i = \frac{\partial x^j}{\partial \bar{x}^i} w_j .$$

The transformation law of geometric objects is quite frequently taken as their definition. We can define a tensor of type (p, q) as an object such that to each chart a system of numbers $T^{i_1 \dots i_p}_{j_1 \dots j_q}$ is assigned, an under a change of the chart, the system transforms in the following way:

$$\bar{T}^{i_1 \dots i_p}_{j_1 \dots j_q} = \frac{\partial \bar{x}^{i_1}}{\partial x^{k_1}} \cdots \frac{\partial \bar{x}^{i_p}}{x^{k_p}} \frac{\partial x^{l_1}}{\partial \bar{x}^{j_1}} \cdots \frac{\partial x^{l_q}}{\partial \bar{x}^{j_q}} T^{k_1 \dots k_p}_{l_1 \dots l_q} .$$

For example, the transformation law for the metric tensor has the form

$$\bar{g}_{ij} = \frac{\partial x^k}{\partial \bar{x}^i} \frac{\partial x^l}{\partial \bar{x}^j} g_{kl} .$$

The Levi-Civitá pseudotensor, the form of which—in inertial systems—we defined in Chapter 8, does not transform in the same way as tensors. Let us require that in any coordinate system ξ the Levi-Civitá pseudotensor ε_{ijkl} be completely antisymmetric (i.e. antisymmetic in any pair of indices) and that $\varepsilon_{0123} = \sqrt{-\det(g_{ij})}$. Since

$$\det(\bar{g}_{ij}) = \left[\det\left(\frac{\partial x^k}{\partial \bar{x}^l}\right)\right]^2 \det(g_{ij}),$$

using the combinational definition of the determinant, we obtain that

$$\bar{\varepsilon}_{ijkl} = \frac{\det(\partial x^k/\partial \bar{x}^l)}{|\det(\partial x^k/\partial \bar{x}^l)|} \frac{\partial x^m}{\partial \bar{x}^i} \frac{\partial x^n}{\partial \bar{x}^j} \frac{\partial x^p}{\partial \bar{x}^k} \frac{\partial x^q}{\partial \bar{x}^l} \varepsilon_{mnpq} .$$

The determinant $\det(g_{ij})$ is an example of a scalar density with weight -2, while ε_{ijkl} is an example of a pseudotensor of type (0, 4). The transformation law for a pseudotensor differs from that for a tensor only in sign when bases are interchanged with a change in orientation. Raising the indices in ε_{ijkl} by means of a metric tensor, we obtain the pseudotensor ε_{ijkl} of type (4, 0) called the Levi-Civitá contravariant pseudotensor. Let us note that $\varepsilon^{0123} = -$ $-(\det(g_{ij}))^{-1/2}$.

Obviously, the transformation laws for tensors, tensor densities and pseudotensors are valid for any differential manifold and not only in an affine space.

We will continue to analyse the structure of the affine space (E, V), attempting to distinguish its "infinitesimal part". We introduce the notion of absolute differentiation in E. Let $q\colon \mathbf{R} \to E$ be a curve in E, and let the vector field $u\colon \mathbf{R} \to V$ (determined only on this curve) be a field tangent to it (Fig. 11.2).

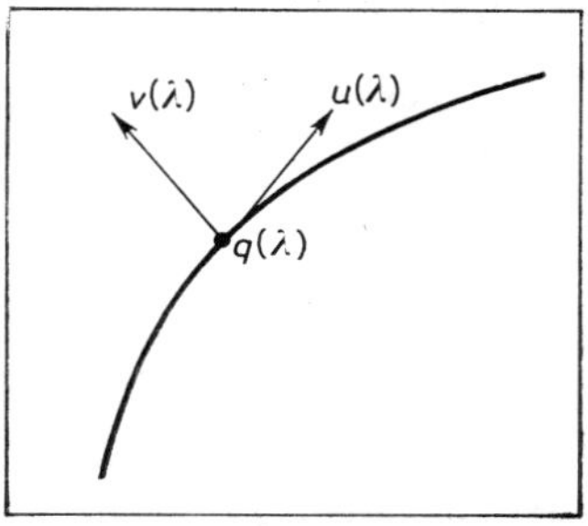

Fig. 11.2

For the vector field $v\colon \mathbf{R} \to V$, determined on this curve, we can introduce the notion of the *absolute derivative* along q

$$\frac{\mathrm{d}v}{\mathrm{d}\lambda} = \lim_{\Delta\lambda \to 0} \frac{v(\lambda+\Delta\lambda)-v(\lambda)}{\Delta\lambda}.$$

This definition is correct, since both vectors $v(\lambda+\Delta\lambda)$ and $v(\lambda)$ belong to the vector space V (this differentiation would not be possible for manifolds without affine space structure). Decomposing the vector v successively with respect to the bases (e_i) and $(\bar{e}_i)$, and then differentiating, we obtain

$$\frac{\mathrm{d}v}{\mathrm{d}\lambda} = \frac{\mathrm{d}v^i}{\mathrm{d}\lambda} e_i = \frac{\mathrm{d}\bar{v}^i}{\mathrm{d}\lambda} \bar{e}_i + \bar{v}^i \frac{\mathrm{d}\bar{e}^i}{\mathrm{d}\lambda}.$$

In the above formula we took into account the fact that the vector fields $\lambda \mapsto e_i$ were constant and the fields $\lambda \mapsto \bar{e}_i$ were variable. Considering the relations between the vectors of the bases,

$$\bar{e}_i = \frac{\partial x^j}{\partial \bar{x}^i} e_j,$$

we obtain that

$$\frac{\mathrm{d}\bar{e}_i}{\mathrm{d}\lambda} = \frac{\partial^2 x^j}{\partial \bar{x}^k \partial \bar{x}^i} \frac{\mathrm{d}\bar{x}^k}{\mathrm{d}\lambda} e_j = \frac{\partial^2 x^j}{\partial \bar{x}^k \partial \bar{x}^i} \frac{\partial \bar{x}^l}{\partial x^j} \bar{u}^k \bar{e}_l .$$

Let us introduce the notation

$$\bar{\Gamma}^l{}_{jk} = \frac{\partial^2 x^j}{\partial \bar{x}^k \partial \bar{x}^i} \frac{\partial \bar{x}^l}{\partial x^j} .$$

We can readily find out that the numbers $\bar{\Gamma}^l{}_{jk}$ are determined by the coordinate system $\bar{\xi}$, but they do not depend on the choice of the rectilinear coordinate system ξ. We interpret $\bar{\Gamma}^l{}_{jk}$ as the lth component (with respect to the basis $(\bar{e}_j)$) of the derivative of the vector $\bar{e}_i$ in the direction of the kth line of the coordinate system $\bar{\xi}$. We can now represent the components of the absolute derivative of the vector field v with respect to the basis $(\bar{e}_j)$, defined by the equation

$$\frac{\mathrm{d}v}{\mathrm{d}\lambda} = \frac{\mathrm{D}\bar{v}^i}{\mathrm{d}\lambda} \bar{e}_i ,$$

by means of Γ:

$$\frac{\mathrm{D}\bar{v}^i}{\mathrm{d}\lambda} = \frac{\mathrm{d}\bar{v}^i}{\mathrm{d}\lambda} + \bar{\Gamma}^i{}_{lk} \bar{v}^l \bar{u}^k .$$

Since the second derivatives of the coordinates x^j over $\bar{x}^i$ appear in the definition of Γ, we can conclude that $\bar{\Gamma}^i{}_{lk} = 0$ if and only if the coordinate system $\bar{\xi}$ is rectilinear. This property of the symbols Γ leads to the statement that they are not tensors since if the coordinates of a tensor are equal to zero in a certain coordinate system, they are equal to zero in any coordinate system. The symbols $\bar{\Gamma}^i{}_{lk}$ form a geometric object, called an *affine connection*, or an object of parallel transport. The latter name comes from the fact that in the language of absolute derivatives the vector v is parallelly transported along the curve q, if

$$\frac{\mathrm{d}v}{\mathrm{d}\lambda} = 0,$$

or, in other words,

$$\frac{\mathrm{D}v^i}{\mathrm{d}\lambda} = 0.$$

We have therefore distinguished in the affine space (E, V) an object of affine connection Γ, which determines absolute differentiation. The affine space is a special case of a space endowed with affine connection, or, more briefly, a space with connection

$$(E, V) \rightarrow (E, \Gamma).$$

The connection Γ is a geometric object which, after a transition from the coordinate system ξ—which is no longer a rectilinear system, since this notion is not well defined in (E, Γ)—to the system $\bar{\xi}$, transforms in the following way:

$$\bar{\Gamma}^i{}_{jk} = \Gamma^l{}_{mn} \frac{\partial \bar{x}^l}{\partial x^l} \frac{\partial x^m}{\partial \bar{x}^j} \frac{\partial x^n}{\partial \bar{x}^k} + \frac{\partial^2 x^l}{\partial \bar{x}^j \partial \bar{x}^k} \frac{\partial \bar{x}^i}{\partial x^l}.$$

We obtain this transformation law in an affine space, and then generalize it to any space with connection.

Our previous considerations suggest that we should take the four-dimensional differential manifold E with the affine connection Γ as a spacetime model which takes into account gravitational phenomena.

In a space with an affine connection, we can, from the equation

$$\frac{\mathrm{D}v^i}{\mathrm{d}\lambda} = \frac{\mathrm{d}v^i}{\mathrm{d}\lambda} + \Gamma^i{}_{jk} v^j u^k,$$

determine the absolute derivative of the vector v along the curve q, and, subsequently, introduce the notion of the parallel transport of vectors.

In a simple and natural way, we can generalize the notion of the absolute derivative (and that of parallel transport) to any tensor fields, making use of the requirements of linearity of this derivative and the demand that it should satisfy the Leibniz formula; for scalar fields, the absolute derivative should be equal to the ordinary derivative.

We then introduce the notion of a covariant derivative. This is a tensor which is obtained by taking absolute derivatives in the directions of coordinate lines. For the lth line, $u^k = \delta^k_l$, and, therefore, the tensor of the covariant derivative of the vector field v^i has the form

$$\nabla_l v^i = \frac{\partial v^i}{\partial x^l} + \Gamma^i{}_{jl} v^j.$$

In order to give the formula for the covariant derivative of any tensor field, let us first consider the covector field w_i. From the Leibniz formula, we have

$$\nabla_l(w_i v^i) = w_i \nabla_l v^i + v^i \nabla_l w_i.$$

Since the covariant derivative of a scalar field is equal to an ordinary derivative

$$v^i \nabla_l w_i = \frac{\partial}{\partial x^l}(w_i v^i) - w_i \frac{\partial v^i}{\partial x^l} - \Gamma^i{}_{jl} w_i v^j = v^i \frac{\partial w_i}{\partial x^l} - \Gamma^j{}_{jl} w_j v^i,$$

and because this formula is valid for any field v^i, we have

$$\nabla_l w_i = \frac{\partial w_i}{\partial x^l} - \Gamma^j{}_{il} w_j.$$

For a simple tensor $v^i w_j$, from the above formulae, using the Leibniz formula, we have

$$\nabla_l(v^i w_j) = \frac{\partial}{\partial x^l}(v^i w_j) + \Gamma^i{}_{kl} v^k w_j - \Gamma^k{}_{jl} w_k.$$

Then, since each tensor is the sum of simple tensors, we have for any tensor of type (1, 1)

$$\nabla_l T^i_j = \frac{\partial}{\partial x^l} T^i_j + \Gamma^i{}_{kl} T^k_j - \Gamma^k{}_{jl} T^i_k.$$

Hence, it is not difficult to guess—and, if there is paper and ink enough, to derive—a formula for the covariant derivative of any tensor field:

$$\nabla_l T^{i_1 \ldots i_p}_{j_1 \ldots j_q}$$

$$= \frac{\partial}{\partial x^l} T^{i_1 \ldots i_p}_{j_1 \ldots}{}_{kl} + \Gamma^{i_1}{}_{kl} T^{k i_2 \ldots i_p}_{j_1 \ldots j_q} + \ldots + \Gamma^{i_p}{}_{kl} T^{i_1 \ldots i_{p-1} k}_{j_1 \ldots j_q}$$

$$- \Gamma^k{}_{j_1 l} T^{i_1 \ldots i_p}_{k j_2 \ldots j_q} - \ldots - \Gamma^k{}_{l_q l} T^{i_1 \ldots i_p}_{j_1 \ldots j_{q-1} k}.$$

The covariant derivative of a tensor (pseudotensor) field of type (p, q) is a tensor (pseudotensor) field of type $(p, q+1)$. (The same formula gives the covariant derivative of a pseudotensor field).

Geodesics are generalizations of the straight lines of affine geometry. Let us bear in mind that a straight line was defined by the equation

$$q(\lambda) = a(\lambda) w + \mathfrak{o},$$

where $w \in V$, $\mathfrak{o} \in E$, while a was a diffeomorphism $a\colon \mathbf{R} \to \mathbf{R}$ (it is enough to require that $a' > 0$). The tangent vector to this straight line is

$$u(\lambda) = a'(\lambda) w;$$

and the derivative of the tangent vector with respect to the parameter λ is proportional to itself:

$$\frac{\mathrm{d}u}{\mathrm{d}\lambda} = a'' w = \frac{a''}{a'} u = bu,$$

or in other words,

$$\frac{\mathrm{D}u^i}{\mathrm{d}\lambda} = bu^i.$$

Let us note, moreover, that the coefficient b is equal to zero if and only if $a(\lambda) = \mu\lambda + \nu$; we say then that λ is an affine parameter.

Geodesics in a space with affine connection are curves satisfying the equation

$$\frac{\mathrm{D}u^i}{\mathrm{d}\lambda} = \frac{\mathrm{d}u^i}{\mathrm{d}\lambda} + \Gamma^i{}_{jk} u^j u^k = bu^i.$$

As we can see, these lines play the same part in spaces with affine connection as straight lines in affine spaces. As in the case when $b = 0$, we say that the parameter λ is affine. We can easily demonstrate that such a parameter always exists, and that if λ is an affine parameter, then only $\mu\lambda+\nu$ is also such a parameter.

In the Newtonian theory of gravitation, the equation of motion for a material point has the form

$$\frac{d^2\mathbf{r}}{dt^2}+\operatorname{grad}\varphi = 0.$$

We can write this equation in another form:

$$\frac{du^i}{dt}+\Gamma^i{}_{jk}u^ju^k = 0,$$

where u is a tangent vector to the world-line of the material point, $(u^i) = (v_x, v_y, v_z, 1)$, while of the affine connection coefficients, only the coefficients $\Gamma^\alpha_{44} = \partial\varphi/\partial x^\alpha$ are different from zero for $\alpha = 1, 2, 3$. We can thus see that the world-lines of material points in the Newtonin theory are geodesics within a certain geometry, and the absolute time t is an affine parameter. The insep arability of the forces of gravitation and inertia is now expressed by the non-tensor character of the coefficients $\Gamma^i{}_{jk}$; in a non-inertial system we have

$$\Gamma^i{}_{jk}u^ju^k \sim \operatorname{grad}\varphi+\text{Coriolis force}+\text{centrifugal force} + \ldots$$

In the geometric approach to the problem of motion in a gravitational field, the question of equality between the gravitational and inertial masses vanishes. Quite simply, in the geodesic equation, there is no room for considering these masses separately.

When is the space with an affine connection, (E, Γ) an affine space? As we have already mentioned, this occurs if there exists a coordinate system such that $\Gamma^i{}_{jk} = 0$. It follows immediately from the above that the quantity,

$$Q^i{}_{jk} = \Gamma^i{}_{kj}-\Gamma^i{}_{jk},$$

which from the transformation law for the coefficients $\Gamma^i{}_{jk}$ is a tensor, must vanish:

$$Q^i{}_{jk} = 0.$$

We call the tensor Q the *torsion tensor*.

The sufficient condition for the space (E, Γ) to be locally affine is the integrability of the equation

$$\frac{\partial^2 x^l}{\partial\bar{x}^j\partial\bar{x}^k} = \Gamma^i{}_{jk}\frac{\partial x^l}{\partial\bar{x}^i},$$

which results from the transformation law for the affine connection coefficietns, if we put $\Gamma^i{}_{jk} = 0$. If the torsion vanishes, the integrability of this equation is ensured by the equality

$$R^i{}_{jkl} = 0,$$

where

$$R^i{}_{jkl} = \frac{\partial \Gamma^i{}_{jk}}{\partial x^l} - \frac{\partial \Gamma^i{}_{jl}}{\partial x^k} + \Gamma^i{}_{ml}\Gamma^m{}_{jk} - \Gamma^i{}_{mk}\Gamma^m{}_{jl}$$

in any coordinate system. The quantity $R^i{}_{jkl}$ is a tensor called the *curvature tensor*. If we add to these two conditions, i.e. the vanishing of the torsion and curvature tensors, a condition stating that whole E is the domain of the chart where the coefficients $\Gamma^i{}_{jk}$ vanish, and moreover that $\mathbf{R}^4$ is the image of this chart, then (E, Γ) will be an affine space.

Calculating the commutator of the covariant derivatives of the vector field, we obtain the formula

$$(\nabla_i \nabla_j - \nabla_j \nabla_i) v^k = R^k{}_{lij} v^l - Q^l{}_{ij} \nabla_l v^k,$$

called the *Ricci identity*. It follows from this formula that in fact $R^k{}_{lij}$ is a tensor. The curvature and torsion tensors are antisymmetric in the last pairs of indices

$$R^i{}_{jkl} = R^i{}_{jkl}, \qquad Q^i{}_{jk} = -Q^i{}_{kj}$$

and they satisfy the differential identities, called the *Bianchi identities*:

$$\nabla_{[j} Q^i{}_{kl]} + Q^i{}_{m[j} Q^m{}_{kl]} = R^i{}_{[jkl]},$$

(where the square brackets mean that we should antisymmetrize all the indices in them), and

$$\nabla_{[k} R^i{}_{|j|lm]} + R^i{}_{jn[k} Q^n{}_{lm]} = 0.$$

(The vertical dashes signify that the index j between them is exempt from antisymmetrization).

Adapting spacetime so that it takes into account the gravitational interactions, we must give up at least the condition that the curvature vanishes. Because we cannot detect torsion by simple mechanical experiments (nor, as it appears, by electromagnetic ones), we usually assume that

$$Q^i{}_{jk} = 0.$$

In the affine spaces of the Galileo theory and the special theory of relativity, we had such metric elements as n, τ and g, which were tensors in V, i.e. constant in E, and therefore satisfied the equations

$$\nabla h = 0, \quad \nabla \tau = 0, \quad \nabla g = 0.$$

In a model of the general theory of relativity, it is thus natural to retain the condition of the vanishing of the covariant derivative of the metric tensor.

The triple (E, Γ, g). where g is a tensor field on E of type (0, 2), satisfying the conditions:

$$Q^i{}_{jk} = 0, \quad \nabla_i g_{jk} = 0,$$

is called the *Riemannian geometry*. Our considerations lead to the suggestion that this geometry is the most appropriate model of spacetime in the relativistic theory of gravitation. It turns out that the conditions given above lead to the conclusion, that the affine connection coefficients Γ^i_{jk}, which, in this case, we call *Christoffel symbols* and denote by $\{^{\,i}_{jk}\}$, are equal to

$$\{^{\,i}_{jk}\} = \frac{1}{2} g^{il}\left(\frac{\partial g_{lj}}{\partial x^k} + \frac{\partial g_{lk}}{\partial x^j} - \frac{\partial g_{jk}}{\partial x^l}\right).$$

In the Riemannian geometry, the connection is determined by the metric, we can therefore say that we have the pair (E, g), which we call the Riemannian geometry.

It follows from the condition $\nabla g = 0$ that the length s, defined by the formula

$$\mathrm{d}s^2 = g_{ij}\mathrm{d}x^i\mathrm{d}x^j,$$

is an affine parameter along the geodesics. Moreover, it turns out that the geodesic equation is equivalent to the variational pinciple

$$\delta\int \mathrm{d}s = 0.$$

Thus, we can see that the geodesics satisfy the necessary condition for the extremal value of the length.

In the Riemannian geometry, we call the curvature tensor a Riemann tensor. The fact that this tensor vanishes over a certain region is equivalent to the possibility of introducing a coordinate system in a neighbourhood of each point of this region such that the coefficients of the metric tensor are constant. The Riemannian spaces, where the Riemann tensor vanished, are called flat. Compared with any curvature tensor, the Riemann tensor has much richer symmetry properties. Let us enumerate them:

$$R^i{}_{jkl} = -R^i{}_{jlk},$$
$$R^i{}_{[jkl]} = 0,$$
$$R_{ijkl} = -R_{jikl},$$
$$R_{ijkl} = R_{klij}.$$

The first of these properties is valid for any curvature tensor. The second is a result of the assumption that torsion vanishes. The third results from the assumption that the covariant derivative of the metric tensor disappears. Finally, the last property results from both conditions imposed on the connection. In an n-dimensional space, a tensor with the symmetries given above has $\frac{n^2}{12}(n^2-1)$ independent components. In the four-dimensional case, this number is 20.

Let us consider which quantity describes the strength of the gravitational field at a given point of spacetime. Although in some region the metric tensor g_{ij} describes fully the gravitational field, an appropriate choice of the coordinate system can bring it at a given point to the standard form from the special theory of relativity. Although the affine connection coefficients give an expression for the gravitational force, they are glued together with the inertial forces. Moreover, at a given point, through an appropriate choice of the system, we can reduce them to zero. The coordinate systems for which at a given point the first derivatives of the metric tensor (and, as a consequence, the connection coefficients) vanish and the metric tensor itself takes the form of the Minkowski matrix, are called local inertial coordinate systems. They are locally the best approximations of the inertial coordinate systems of the special relativity theory. In these systems, the physical laws determined by the special relativity theory are valid locally.

It is only the Riemann tensor containing the second derivatives of the components of the metric tensor that we can regard as the strength of the gravitational field. In contrast to electromagnetism, this strength is not related to a force, so it does not provide information about the motion of a single particle. It contains, on the other hand, information about the motions of adjacent particles. For if we consider a family of the world-lines of particles, namely the geodesics $x^i(s, \eta)$ in the neighbourhood of the standard geodesic $x^i(s) = x^i(s, \eta = 0)$ and determine the separation four-vector $n^i = \left.\dfrac{\partial x^i}{\partial \eta}\right|_{\eta=0}$, the simple calculations involving the use of the Ricci identity give, the so-called geodesic deviation equation

$$\frac{\mathrm{D}^2}{\mathrm{d}s^2} n^i = R^i{}_{jkl} u^j u^k n^l.$$

Therefore, measuring the relative acceleration of adjacent test particles in the gravitational field, we can, in principle, determine the curvature tensor.

Taking the trace from the Riemann tensor, we obtain the Ricci tensor

$$R_{ij} = R^k{}_{ikj},$$

which is symmetric, $R_{ij} = R_{ji}$. Taking the trace once again, we obtain the Ricci curvature scalar

$$R = R_{ij} g^{ij}.$$

In two-dimensional space the Riemann tensor is expressed by the curvature scalar. In three-dimensional space, in turn, the Riemann tensor is expressed by the Ricci tensor. In space with dimensions $n > 3$, the number of the independent components of the Riemann tensor exceeds that of the independent components of the Ricci tensor, and, accordingly, the Ricci tensor does not contain all information about the Riemann tensor. We can then introduce the Weyl tensor $C^i{}_{jkl}$ describing this information in the Riemann tensor that is not contained in the Ricci tensor. It is defined by the following formula:

$$C^{ij}{}_{kl} = R^{ij}{}_{kl} + \frac{4}{n-2}\,\delta^{[i}_{[k} R^{j]}_{l]} - \frac{2}{(n-1)(n-2)}\,\delta^{i}_{[k}\delta^{j}_{l]}\,R.$$

This tensor has all these properties of symmetry that the Riemann tensor has, and, moreover, it is tracefree in any pair of indices. In four-dimensional space, the number of independent components is 10. In a tensor space with the symmetries of the Weyl tensor, the orthogonal group acts in an irreducible way, and, therefore, this tensor cannot be further decomposed.

We call the operations

$$g_{ij} \to g'_{ij} = \mathrm{e}^{2U} g_{ij},$$

where U is an arbitrary function, conformal transformations of the metric tensor. For these transformations, the Weyl tensor does not change, $C'^i{}_{jkl} = C^i{}_{jkl}$. Therefore, quite frequently, it is also called the *conformal curvature tensor*. In particular, this tensor vanishes if and only if the metric is conformally flat, i.e. if it is locally proportional to the flat space metric.

In a three-dimensional space the Weyl tensor vanishes identically. This does not mean, however, that any three-dimensional space is conformally flat. The condition for conformal flatness is here the equation

$$\nabla_{[i} R^k_{j]} - \frac{1}{4}\nabla_{[i} R \delta^k_{j]} = 0.$$

A two-dimensional space is, on the other hand, always conformally flat.

CHAPTER 12

Einstein Equations

We now limit our considerations to a four-dimensional Riemannian space (E, Γ, g). Just as in the special theory of relativity, we assume that the metric tensor g has the signature $(+, -, -, -)$.

Let us now give a scheme which shows the relations between the theories of spacetime discussed here.

Galileo theory E, Γ, h, τ $Q = 0,\ R = 0,\ \nabla h = 0,$ $\nabla \tau = 0$	**SRT** E, Γ, g $Q = 0,\ R = 0,\ \nabla g = 0$
Newton theory E, Γ, h, τ $Q = 0,\ \nabla h = 0,\ \nabla \tau = 0$ $\mathrm{Tr}\, R \sim \varrho \leftrightarrow \Delta\varphi = 4\pi k \varrho$ free falls $\leftrightarrow$ geodesics	**GRT** E, Γ, g $Q = 0,\ \nabla g = 0$ Gravitational field equations? free falls $\leftrightarrow$ timelike geodesics light rays $\leftrightarrow$ null geodesics ideal clocks measure s

In this scheme, SRT and GRT denote the special and general relativity theories, respectively. $\mathrm{Tr}\, R$ is the *Ricci tensor*.

We should note here that Einstein did not build the theory of gravitation on the Newtonian level, but set out from the special theory of relativity. Perhaps, if Einstein had preceded the construction of the general theory of relativity by a geometrization of the Newtonian theory, he would have reached his goal earlier than he did. If we neglect the problem of the form of gravitational field equations, the formulation of the GRT given above is contained in a paper which Einstein wrote in 1913 together with the mathematician M. Grossmann [12]. The field equations on the other hand have a long and complicated history.

How should we write the gravitational field equations? So far, we know only that in the nonrelativistic limit these equations should turn into the equations of the Newtonian theory of gravitation. In this theory, on the right-hand side of the equation we have the mass density. What can we expect on the right-hand side of the GRT equations?

Relativistically, mass corresponds to energy, and the mass density to energy density, which, in turn, is proportional to T_{00}—a component of the energy momentum tensor. Hence the idea of considering the energy-momentum tensor T_{ij} as the source of the gravitational field. Since the coefficients of affine connection, which are linear combinations of the derivatives of a metric tensor, were in the Newtonian theory proportional to the first derivatives of the potential φ, we can suppose that we should regard g_{ij} as gravitational potentials.

The natural generalization of the Laplacian occurring in the Newtonian theory of gravitation is of course the d'Alembertian, and the left side of the gravitational field equations should therefore contain $\Box g_{ij}$. Moreover, we should add a term accounting for the energy and momentum of the gravitational field itself to the energy-momentum density tensor; by analogy with electrodynamics, this term should be a quadratic function of the derivatives of the metric tensor. Finally, we can symbolically represent the gravitational field equations proposed by Einstein in 1913 and 1914 [12, 13] as

$$\Box g_{ij} = \varkappa T_{ij} + (\partial g/\partial x)^2_{ij}.$$

Because of the occurrence of first derivatives (and also the d'Alembertian), these equations are not tensorial; in fact, it turns out that we cannot construct any tensor from the first derivatives of the metric tensor. This means that these equations are satisfied only in certain coordinate systems. At first, on the grounds of causality, Einstein rejected the tensor form of the equations, which we can write symbolically as

$$\text{tensor}(g_{ij}) = \varkappa T_{ij}.$$

Since these equations are tensorial, any transformation of coordinates transforms the solutions of these equations into some other solutions.

Let us now suppose that we have Cauchy data, given on the hypersurface $x^0 = 0$. Since the gravitational field equations are hyperbolic, the initial data should determine such quantities as the density of matter or the (Ricci) curvature scalar at future-oriented points. Let, e.g. the point (0, 0, 0, 0) lie on the same world-line of matter as point (1, 0, 0, 0). Let us now perform a transformation of the coordinates which is reduced to an identity in the neighbourhood of the surface $x^0 = 0$. In this way, we obtain a different solution of the same field equations, but now the point (1, 0, 0, 0), which still lies on the same world-line as point (0, 0, 0, 0), is in another place (Fig. 12.1), while, e.g. the

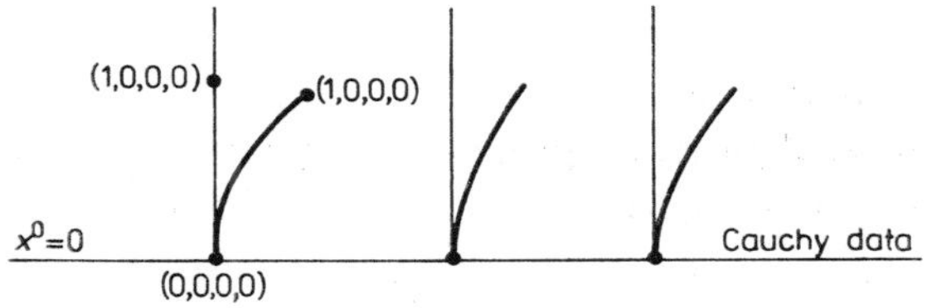

Fig. 12.1

density of matter in the previous point will in general be different. Thus, on the basis of tensor equations, we cannot predict what the density of matter will be, say in an hour. "Obviously", this disqualifies [13] these equations.

The above train of thought is not valid in view of the fact that the coordicenates themselves are not physically meaningful, but that only a point in space-time is. After any transformation of the coordinates, the value of the energy density and the curvature scalar does not change; while the fact that the same point now has different coordinates in a new system should not really surprise anyone. Einstein used this mistaken argumentation until 1915 when he recognized that, on the contrary, the field equation should have a generally invariant, i.e. tensor, character [14, 16]. Assuming this point of view, we come to the conclusion that singled-out systems of reference do not exist, and that the group of automorphisms of GRT is the group of all transformations of coordinates.

In 1915 Einstein proposed [14] the gravitational equations

$$R_{ij} = \varkappa T_{ij}.$$

Because of the generalized law of conservation of the energy-momentum $\nabla_j T^{ij} = 0$, this equation imposes an additional differential condition on the Ricci tensor and causes the overdeterminacy of the system of gravitational field equations. For this reason, in late 1915 [15], Einstein replaced it with the equation

$$G_{ij} = \varkappa T_{ij}$$

where $G_{ij} = R_{ij} - \frac{1}{2} g_{ij} R$ is called the *Einstein tensor*. The equation $\nabla_j T^{ij} = 0$ is now simply a conclusion from the so-called *contracted Bianchi identity*

$$\nabla_j G^{ij} \equiv 0.$$

This situation resembles that in electrodynamics. Here, from the Maxwell equations

$$\nabla_j F^{ij} = -\frac{4\pi}{c} j^i,$$

after taking into account the identity

$$\nabla_i \nabla_j F^{ij} = 0,$$

resulting from the antisymmetry of the tensor of the electromagnetic field F^{ij}, we obtain the equation of continuity of the current density vector

$$\nabla_k j^k = 0,$$

i.e. the law of conservation of charge.

As all fundamental physical equations the equations $G_{ij} = \varkappa T_{ij}$ result from a certain variational principle. In this case, it has the form

$$\delta \int \left(\mathscr{L} - \frac{1}{2\varkappa} R \right) d\omega = 0,$$

where $d\omega$ is the density four-form and $\mathscr{L}$ is an invariant Lagrangian of matter. We carry out the variation with respect to g_{ij}.

We still have to determine the value of the coefficient $\varkappa$. To find it, we shall consider a weak gravitational field with corresponding small curvatures, and we shall deal with not very fast motions, that is, $v/c \ll 1$. Then, in the equation of geodesics

$$\frac{d^2x^i}{ds^2}+\Gamma^i{}_{jk}\frac{dx^j}{ds}\frac{dx^k}{ds}=0.$$

we can make a number of approximations. First, we put

$$ds \cong c\,dt,$$

hence, it follows that

$$\left(\frac{dx^i}{ds}\right) \cong \left(1, \frac{\mathbf{v}}{c}\right).$$

Then, neglecting the first-order terms in v/c, we have

$$\Gamma^i{}_{jk}\frac{dx^j}{ds}\frac{dx^k}{ds} \cong \Gamma^i{}_{00} = \frac{1}{2}g^{ij}\left(2\frac{\partial g_{j0}}{\partial x^0}-\frac{\partial g_{00}}{\partial x^j}\right).$$

Since the motions are slow, the ratio of the time derivatives to the spatial derivatives is of the order of $v/c \ll 1$, and, taking into account the weakness of the gravitational field

$$g_{ij} = \eta_{ij}+\text{small corrections},$$

we have no difficulty in obtaining that

$$\Gamma^i{}_{00} \cong \frac{1}{2}\frac{\partial g_{00}}{\partial x^i},$$

and, thus, the spatial components of the equation of geodesics take the approximate form

$$\frac{d^2x^\alpha}{dt^2}+c^2\frac{1}{2}\frac{\partial g_{00}}{\partial x^\alpha}=0 \qquad \text{for } \alpha = 1, 2, 3.$$

By comparison with the classical Newtonian equation

$$\frac{d^2x^\alpha}{dt^2}+\frac{\partial\varphi}{\partial x^\alpha}=0,$$

we obtain the approximate result

$$g_{00} \cong 1+\frac{2\varphi}{c^2}.$$

In the case of perfect dust

$$T_{ij} = \varrho c^2 u_i u_j,$$

the Einstein equation

$$G_{ij} = \varkappa T_{ij}$$

gives

$$\Delta\varphi \cong \frac{1}{2}\varkappa c^4 \varrho\,.$$

By comparison with the classic Poisson's equation

$$\Delta\varphi = 4\pi k \varrho$$

we finally obtain

$$\varkappa = 8\pi k/c^4.$$

CHAPTER 13

Some Aspects of the General Relativity Theory

The general relativity theory formulated in the previous chapter does not single out any reference systems. The equations of this theory have a tensorial form. If we have the solution of these equations in a certain coordinate system and if we then transform it into another system, the transformed solution satisfies the initial equation expressed in the new coordinate system. The above property of the general relativity theory is often called the principle of general covariance. We can formulate it otherwise in the following way: the group of symmetries of the general relativity theory is the full group of diffeomorphisms of spacetime.

What do we call the *group of symmetries of a physical theory*? In every theory, we can distinguish *absolute* and *dynamical elements*. The latter serve to describe the history of a physical system and, in contrast to the former, change in the course of motion. The group of automorphisms of the absolute elements of a physical theory is called the group of symmetries of this theory. In a few examples we shall show how the dividing line between absolute and dynamical elements runs, and what groups of symmetries look like.

The first example is Newtonian mechanics where we have the following elements:

$$(\underbrace{E, V, h, \tau;}_{\text{absolute}} \quad \underbrace{\text{coordinates of particles, densities, pressures, ...}}_{\text{dynamical}}),$$

The Galilean group is here the group of automorphisms of the absolute elements.

Another, more particular problem is the classical one-body problem. Apart from the absolute elements given above, two other such elements are involved here, namely the system in which the body is at rest (the ether of this body e) and the force with which the body acts on another one. The force is characterized by a potential, e.g. by the gravitational potential $V = \alpha/r$:

$$(\underbrace{E, V, h, \tau, V = \alpha/r, e;}_{\text{absolute}} \quad \underbrace{\text{coordinates of the particle}}_{\text{dynamical}}).$$

In the one-body problem the group of automorphisms consists of the group of rotations in three-dimensional space and of time translations.

In the special relativity theory we have a situation resembling that in Newtonian mechanics:

$$(\underbrace{E, V, g}_{\text{absolute}}; \underbrace{\text{world-lines of points, electromagnetic field, ...}}_{\text{dynamical}}),$$

the group of symmetries is of course the Poincaré group.

In quantum theories, the situation hardly changes. Let us consider the quantum theory of a certain system characterized by the Hilbert space $\mathscr{H}$ and the Hamiltonian operator $\hat{H}$. Here, we have the following elements:

$$(\underbrace{\mathscr{H}, \hat{H}}_{\text{absolute}}; \underbrace{\text{history of the system's states}}_{\text{dynamical}}).$$

Now, the group of automorphisms is the set of unitary operators in the space $\mathscr{H}$ which commute with the Hamiltonian operator.

In all the above-mentioned cases, the group of symmetries is a Lie group, and, thus, the actions are continuous and these groups have a finite dimension. The situation is different in the general relativity theory:

$$(E; g, \text{electromagnetic field, matter velocity field, ...}).$$

Here, the only absolute element is the "bare" spacetime E itself. The scalar product g is subject to the Einstein equation, and is thus not an absolute element.

The group of symmetries is the full group of automorphisms of the manifold E, which is not a Lie group. Some physicists, notably Fock, consider this situation unsatisfactory. Fock proposes that in spacetime another plane metric η should be introduced as an absolute element:

$$(\underbrace{E, \eta}_{\text{absolute}}; \underbrace{g, \text{as above}}_{\text{dynamical}}).$$

The scalar product η is to be determined by singling out a certain class of systems which are called harmonic and function as inertial systems. It appears that such a bi-metric theory is not so good, for it is difficult to give a physical interpretation of the metric η (or, which is really the same thing, that of the corresponding inertial systems).

Let us now deal with the consequences of the presence of symmetries and general invariance. In this connection, let us introduce the notions of the differential form and the Lie derivative.

Differential forms play a particular role among tensor fields, since we can differentiate them without resorting to the notion of an affine connection. It is worth noting that we can fully formulate the Maxwell differential equations in the language of differential forms.

The *differential form* ω *of degree* p is an antisymmetric tensor field of type $(0, p)$. The antisymmetry means that the coordinates of the tensor field ω change sign if we transpose any pair of indices, e.g. for the form of degree 2

$$\omega_{ij} = -\omega_{ij}.$$

When considered in one point of an n-dimensional manifold, differential forms of degree p form a vector space of dimension $\binom{n}{p}$. The external product of the forms $\overset{p}{\omega}$ and $\overset{q}{\omega}$ of degrees p and q respectively is a form of degree $p+q$ whose components are given by the formula

$$(\overset{p}{\omega} \wedge \overset{q}{\omega})_{i_1 \ldots ip+q} = \frac{1}{(p+q)!} \sum_{\sigma} \pm \overset{p}{\omega}_{\sigma(i_1)\ldots\sigma(i_p)} \overset{q}{\omega}_{\sigma(i_{p+1})\ldots\sigma(ip+q)},$$

where the summation extends over all the permutations σ, and the $+$ or $-$ sign depends on whether a given permutation is even or odd. The external multiplication is linear with respect to both arguments; it is also associative. In general, it is not commutative, but we have

$$\overset{p}{\omega} \wedge \overset{q}{\omega} = (-1)^{pq} \overset{q}{\omega} \wedge \overset{p}{\omega}.$$

If (x^i) is a coordinate system on the manifold, forms of degree 1, dx^i span the space of forms of degree 1 at each point, while their external products provide the basis for any differential forms, i.e. each differential form has the form

$$\omega = \omega_{i_1 \ldots i_p} dx^{i_1} \wedge \ldots \wedge dx^{i_p}.$$

If ω is a form of degree $p > 0$, we can determine its *internal product* with the vector field X, which is a form of degree $p-1$ given by the formula

$$X \lrcorner \omega = p X^{i_1} \omega_{i_1 i_2 \ldots i_p} dx^{i_2} \wedge \ldots \wedge dx^{i_p}.$$

We define the operation of external differentiation of differential forms by the formula

$$d\omega = d\omega_{i_1 \ldots i_p} \wedge dx^{i_1} \wedge \ldots \wedge dx^{i_p}.$$

We can see that this operation is a generalization of the differential of a function.

Let E be a differential manifold (later—spacetime) and let there be given the differentiable mapping

$$\mathbf{R} \times E \ni (t, p) \mapsto \varphi_t(p) \in E,$$

satisfying the conditions

$$\varphi_t \circ \varphi_s = \varphi_{t+s}, \qquad \varphi_0 = \mathrm{id}_E.$$

The set of functions $\{\varphi_t : t \in \mathbf{R}\}$ satisfying these conditions is called a *one-parameter group of transformations of E.* We can readily notice that

$$\varphi_t^{-1} = \varphi_{-t},$$

and, hence, we can see that this set is a subgroup of the group of diffeomorphisms of the manifold E.

Given a certain one-parameter group of transformations of E, we can transport functions, vector fields, fields of differential forms and other tensor fields onto E. If $f: E \to \mathbf{R}$ we define the transported function by the formula

$$f_t = f \circ \varphi_t.$$

We transport the differentials of the function according to the equation

$$(\mathrm{d}f)_t = \mathrm{d}(f_t),$$

and then we generalize this definition to any differential forms. To transfer the vector field Y, we use the equation

$$Y_t \lrcorner \mathrm{d}f_t = Y \lrcorner \mathrm{d}f.$$

It is not difficult to generalize the notion of transport to any tensor field, but we shall not describe it here.

Through any point $p \in E$ we can trace a curve $t \mapsto p(t)$, defined by the equation

$$p(t) = \varphi_t(p).$$

This curve is called the *orbit* of the point, p generated by the one-parameter group of transformations, while the set of the orbits of all the points of E consists of the trajectories of this group. Exactly one trajectory runs through each point $p \in E$. The vec tor field X, tangent to all the trajectories of the group is called the *field induced by this group* (Fig. 13.1).

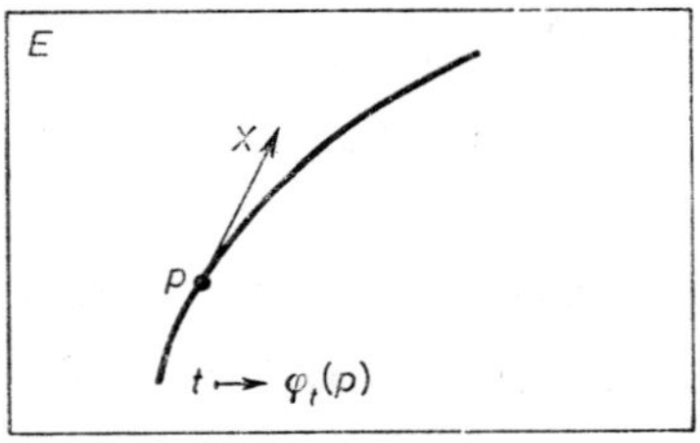

Fig. 13.1

We call the following expression the *Lie derivative* of the tensor field ψ with respect to X:

$$\underset{X}{\pounds}\psi = \frac{\mathrm{d}}{\mathrm{d}t}\,\psi_t|_{t=0},$$

which is a tensor field of the same type as that of the initial field ψ. The vanishing of the Lie derivative of the field ψ,

$$\underset{X}{\pounds}\psi = 0,$$

is equivalent to the constancy of the field ψ on the trajectories of the group:

$$\psi_t = \psi$$

for any $t \in \mathbf{R}$.

Let us note that for any differential form we have

$$\underset{X}{\pounds}\omega = \mathrm{d}(X \rfloor \omega) + X \rfloor \mathrm{d}\omega .$$

Let there be given a certain physical theory $(E, \gamma; \psi)$ where E is spacetime, γ describes absolute elements different from E and ψ describes dynamical elements. Let the dynamical equations of this theory result from the variational principle

$$\delta \int_{\Omega} \lambda = 0,$$

where we assume that the variations ψ vanish on the boundary of the region Ω:

$$\delta\psi|_{\partial\Omega} = 0 .$$

From this variational principle, there follow the field equations of the quantity ψ, which we represent symbolically as $\Psi = 0$.

The four-form λ depends on ψ and γ. One proves the general identity

$$\underset{X}{\pounds}\lambda = \Gamma \underset{X}{\pounds}\gamma + \Psi \underset{X}{\pounds}\psi + \mathrm{d}\chi ,$$

where χ is a three-form. Moreover, because λ is a four-form, and therefore $\mathrm{d}\lambda = 0$, we have

$$\underset{X}{\pounds}\lambda = \mathrm{d}(X \rfloor \lambda) .$$

If the absolute elements γ are constant along the trajectory of the field X:

$$\underset{X}{\pounds}\gamma = 0,$$

and the dynamic equations

$$\Psi = 0,$$

are satisfied, the conservation law of the quantity $\chi - X \rfloor \lambda$ is valid:

$$\mathrm{d}(\chi - X \rfloor \lambda) = 0 .$$

Usually, the equation $\underset{X}{\pounds}\gamma = 0$ determines a certain finite-dimensional Lie algebra of vector fields on E which are tangent to the trajectories of the group of automorphisms of the absolute elements, i.e. to the group of symmetries of a given physical theory. A certain conservation law corresponds to each group of symmetries.

A particular situation arises in the relativistic theory of gravitation where, apart from E, there are no absolute elements, and the equation $\underset{X}{\pounds}\gamma = 0$ is satisfied by any vector field X, corresponding to the principle of general invariance. In this case, we have the identity

$$\Psi \underset{X}{\pounds}\psi = \mathrm{d}(X \lrcorner \lambda - \chi).$$

If the field X vanishes on the boundary of the region Ω,

$$X|_{\partial\Omega} = 0,$$

then χ vanishes there, too, and using the Stokes theorem we obtain that

$$\int_{\Omega} \Psi \underset{X}{\pounds}\psi = 0,$$

which leads to the so-called generalized Bianchi identities. For example, in the theory of gravitation they have the form

$$\nabla_j G^{ij} = 0,$$

and in electrodynamics,

$$\nabla_j(\nabla_k F^{kj}) = 0.$$

The logical structure of the general relativity theory is slightly different from that of other classical field theories. Let us take, e.g., electrodynamics. Here, from the Maxwell equations and from the equations of motion, there follow the conservation laws of energy and momentum given by the equation

$$\nabla_j T^{ij} = 0.$$

The case is different in the theory of gravitation. Here, from the Einstein equations,

$$G^{ij} = \varkappa T^{ij},$$

by taking into account the identity $\nabla_j G^{ij} \equiv 0$ we obtain the conservation laws

$$\nabla_j T^{ij} = 0,$$

and hence, the equation of motion of particles.

Let us show, as an example, how the equations of motion result from the Einstein equations for perfect dust, i.e. for a system of non-interacting particles, whose energy-momentum tensor has the form

$$T^{ij} = c^2 \varrho u^i u^j,$$

where ϱ is the density of dust and u^i are the components of the four-velocity field. The conservation law of energy and momentum, resulting from the Einstein equations has the form

$$0 = \nabla_j(\varrho u^i u^j) = u^i \nabla_j(\varrho u^j) + \varrho u^j \nabla_j u^i.$$

Let us note that the second term in this formula contains a factor equal to the absolute derivative of the four-velocity along the world-line of dust

$$u^j \nabla_j u^i = \mathrm{D}u^i/\mathrm{d}s.$$

Since the velocity four-vector is normalized

$$u_i u^i = 1,$$

differentiating this equality along the world-line of dust, w obtain

$$u_i \frac{\mathrm{D}u^i}{\mathrm{d}s} = 0.$$

Multiplying the equation expressing the conservation law of energy and momentum by u^i, we obtain

$$\nabla_j(\varrho u^j) = 0,$$

i.e. the continuity equation expressing the conservation law of mass. Moreover, returning to the initial equation, we have

$$\frac{\mathrm{D}u^i}{\mathrm{d}s} = 0,$$

i.e. the equation of free particle motion.

In electrodynamics, we can propose the form of the electromagnetic field and find the motion of charges in this arbitrarily defined field. We can also do the opposite: having a given motion of charges, we can determine the field. This is possible because the equations of motion and the field equations are independent. In the general relativity theory, on the other hand, it is impossible to do so, because the equations of motion result from the field equations. This causes complications: we have to solve the motion and field equations simultaneously. In the general case, we cannot solve the problem of motion exactly. Einstein, Infeld and Hoffmann, and, independently, Fock, gave an approximate procedure for dermining the world-lines of matter, consisting of expanding all the functions occurring in the field equations with respect to negative powers of the speed of light.

CHAPTER 14

Algebraic Classification of Gravitational Fields

In Chapter 11 we said that the Riemann tensor plays the role of the intensity of the gravitational field. Therefore, the algebraic classification of gravitational fields is concerned with the Riemann tensor. A full classification should concern both the Ricci tensor and the Weyl tensor. Moreover, since the Ricci tensor is directly related to the sources of the gravitational field, its classification in fact concerns the sources, namely the energy-momentum tensor. Therefore, in the first part of this chapter, we shall classify the energy-momentum tensor, and in the second part, we shall deal with the proper classification of gravitational fields, i.e. the classification of the Weyl tensor.

We shall denote the energy-momentum tensor by T and consider it a linear mapping of the tangent space V at a certain chosen point of spacetime into itself. That is, if $u = (u^i) \in V$, then Tu denotes a vector from V with the components $T^i_j u^j$. The condition of symmetry of the energy-momentum tensor takes the form

$$g(Tu, v) = g(u, Tv)$$

for any $u, v \in V$.

Let us bear in mind that we call the vector $u \neq 0$ an eigenvector corresponding to the eigenvalue $\lambda \in \mathbf{R}$, if

$$Tu = \lambda u.$$

The eigenvector exists if and only if λ is the solution of the characteristic equation

$$\det(T - \lambda \,\mathrm{id}) = 0.$$

If a complex solution of this equation exists, there are no (real) eigenvectors; however, complex eigenvectors, i.e. eigenvectors belonging to $V^{\mathbf{C}}$, do exist. We say that the subspace $W \subset V$ is invariant (for the operator T), if $TW \subset W$. We call the invariant subspace $W_\lambda = \{u \in V : Tu = \lambda u\}$ an eigenspace corresponding to the eigenvalue λ.

If all solutions of the characteristic equation are real, then there exists a basis in V such that the tensor T takes the canonical Jordan form. This is the block-diagonal form, where the diagonal is made up of so-called *Jordan cages*, i.e. matrices of the form

$$\begin{pmatrix} \lambda & 1 & & 0 \\ & \ddots & \ddots & \\ & & \ddots & 1 \\ 0 & & & \lambda \end{pmatrix}.$$

In Segré notation, the dimensions of successive Jordan cages are usd to determine the type of linear transformation. For example, type [2, 1, 1] denotes the following canonical Jordan form:

$$\begin{pmatrix} \lambda_1 & 1 & 0 & 0 \\ 0 & \lambda_1 & 0 & 0 \\ 0 & 0 & \lambda_2 & 0 \\ 0 & 0 & 0 & \lambda_3 \end{pmatrix}.$$

In the case where the eigenvalues corresponding to the different cages are equal, we speak of degeneration. To mark this fact, we put the numbers representing the dimensions of the appropriate Jordan cages in round brackets. In the above example, if $\lambda_1 = \lambda_2 \neq \lambda_3$, we have the Segré type [(2, 1)1]. If complex solutions of the characteristic equation (complex eigenvalues) exist, then there is a basis in $V^{\mathbf{C}}$ such that the tensor T (extended in a natural way to $V^{\mathbf{C}}$) takes the canonical Jordan form. We can then also use the Segré otation, complementing the symbol of the Jordan cage with the letter z, if the given eigenvalue is complex. Since the tensor T is a real tensor, the complex eigenvalues always occur in pairs: λ, $\bar{\lambda}$.

We know perfectly well that in the case of Euclidean space, a symmetrical linear transformation is of the diagonal type [1, 1, 1, 1], or is one of its degenerations: [(1, 1) 1, 1], [(1, 1) (1, 1)], [(1, 1, 1) 1] and [(1, 1, 1, 1)]. In the case of Minkowski space the situation is more complex. Nevertheless, a number of facts known in the Euclidean case are still valid. They are given by

Theorem 1

If W is an invariant subspace, the subspace

$$W_\perp = \{u \in V: g(u, v) = 0 \text{ for all } v \in W\}$$

is invariant. If W is an invariant spacelike subspace, there exists an orthonormal basis in W such that $T|_W$ is diagonal.

We leave the proof to the reader. We still need

Theorem 2

There exist invariant subspaces of arbitrary dimension.

Proof: In the case of real eigenvalues only, the thesis follows from the canonical Jordan form. If there exists a complex eigenvalue, the real and imaginary parts of the corresponding eigenvector span the two-dimensional eigenspace W. The subspace W cannot be spacelike, since in this case the symmetric transformation T_W is diagonal, and therefore it has only real eigenvalues. The subspace W cannot be null, either, since then $W \cap W^{\perp}$ would be a one-dimensional null eigenspace, and then there would exist in W a real eigenvector and the corresponding real eigenvalue. Therefore, the subspace W is timelike. In the invariant spacelike subspace $W^{\perp}$, there exist one-dimensional invariant subspaces W_1 and W_2, generated by eigenvectors; $W+W_1$ and $W+W_2$ are three-dimensional invariant subspaces.

From these theorems, we can include the symmetrical tensor T within one of four classes: I, C, II and III. Here are their characteristics:

I. There exists a timelike eigenvector

If such a vector exists, the subspace orthogonal to it is an invariant three-dimensional spacelike subspace where the linear transformation T is diagonal. There exists, then, the orthonormal basis (e_0, e_1, e_2, e_3) in V such that T has the following canonical form:

$$(T^i_j) = \begin{pmatrix} \lambda_0 & & & 0 \\ & \lambda_1 & & \\ & & \lambda_2 & \\ 0 & & & \lambda_3 \end{pmatrix},$$

or

$$T^{ij} = \lambda_0 e^i_0 e^j_0 - \lambda_1 e^i_1 e^j_1 - \lambda_2 e^i_2 e^i_2 - \lambda_3 e^i_3 e^i_3 .$$

C. There exists an invariant two-dimensional timelike subspace without (real) eigenvectors

From the proof of Theorem 2, such a situation occurs if and only if T has a complex eigenvalue. In the invariant two-dimensional timelike subspace W, we shall choose a basis (k, l) of W made up from null vectors, such that $g(k, l) = 1$. In this basis

$$Tk = \alpha k + \beta l,$$

$$Tl = \gamma k + \delta l .$$

The symmetry condition $g(k, Tl) = g(l, Tk)$ gives $\delta = \alpha$.
For $T|_w$ the characteristic equation has the form

$$\lambda^2 - 2\alpha\lambda + \alpha^2 - \beta\gamma = 0 .$$

A real solution does not exist if $\beta\gamma < 0$. Using the Lorentz transformation $(k \mapsto e^{\varphi} k, l \mapsto e^{-\varphi} l)$, we can achieve $\gamma = -\beta$. Taking the orthonormal eigenvectors (e_1, e_2) from the subspace $W^{\perp}$, we obtain a basis (k, l, e_1, e_2) of V such that, finally,

$$T^i_j = \begin{pmatrix} \alpha & -\beta & & 0 \\ \beta & \alpha & & \\ & & \lambda_1 & \\ 0 & & & \lambda_2 \end{pmatrix},$$

or

$$T^{ij} = \alpha(k^i k^j + l^i l^j) + \beta(l^i l^j - k^i k^j) - \alpha_1 e_1^i e_1^j - \lambda_2 e_2^i e_2^j,$$

where $\beta \neq 0$. The complex eigenvalues are the numbers $\lambda = \alpha + \mathrm{i}\beta$ and $\bar{\lambda} = \alpha - \mathrm{i}\beta$, while the corresponding eigenvectors have the forms $u = k - il$ and $\bar{u} = k + \mathrm{i}l$.

II. There exists an invariant two-dimensional timelike subspace with a non-timelike eigenvector

Let us denote this subspace by W and this eigenvector (necessarily null) as k. In the basis (k, l) of the subspace W, constructed as in the class C, taking into account the symmetry conditions, we have

$$Tk = \lambda_0 k,$$
$$Tl = \alpha k + \lambda_0 l,$$

where $\alpha \neq 0$, since otherwise we find ourselves in class I. Using the Lorentz transformation $(k \mapsto \mathrm{e}^{\varphi} k,\ l \mapsto \mathrm{e}^{-\varphi}\, l)$, we can achieve $\alpha = \pm 1$. The sign of α is an invariant of Lorentz transformations; thus, depending on its value, we have subclasses II_+ and II_-. In the basis $(k, 1, e_1, e_2)$ of V, with orthogonality properties such as in class C, we have

$$(T^i_j) = \begin{pmatrix} \lambda_0 & \pm 1 & & \\ & \lambda_0 & & 0 \\ & & \lambda_1 & \\ & 0 & & \lambda_2 \end{pmatrix},$$

or

$$T^{ij} = \pm k^i k^j + \lambda_0(k^i l^j + l^i k^j) - \lambda_1 e_1^i e_1^j - \lambda_2 e_2^i e_2^j.$$

III. There is no invariant two-dimensional timelike subspace

In this case, we have the pair W, $W^\perp$ of invariant two-dimensional null subspaces. The null eigenvector $k \in W \cap W^\perp$, together with the normalized spacelike vectors e_1 and e_2, span the subspaces $W^\perp$ and W, respectively. We have

$$Tk = \lambda_0 k,$$
$$Te_1 = \alpha k + \lambda_1 e_1,$$
$$Te_2 = \beta k + \lambda_2 e_2.$$

Both numbers λ_1 and λ_2 cannot be different from λ_0, since then spacelike eigenvectors would exist in both subspaces W and $W^\perp$ and the subspace orthogonal to them would be two-dimensional, timelike and invariant. We can therefore take $\lambda_0 = \lambda_1$. Now if $\lambda_2 \neq \lambda_1$, then a spacelike eigenvector exists in the plane W. If, on the other hand, $\lambda_2 = \lambda_1$, then a spacelike eigenvector exists which is a combination of e_1 and e_2. In both cases we can assume that $\beta = 0$. Since $\alpha \neq 0$, rescaling the vector k we can obtain $\alpha = 1$. Choosing the vector l as in the previous cases, we obtain from the symmetry conditions,

$$Tl = 2\gamma k + \lambda_1 l - e_1 .$$

Performing the Lorentz transformation

$$e_1 \mapsto e_1 - \gamma k,$$

$$l \mapsto l + \frac{\gamma^2}{2} k - \gamma e_1,$$

we eliminate the coefficient γ; $\gamma = 0$. In the basis (k, e_1, l, e_2) of V we thus obtain the following canonical form:

$$(T^i_j) = \begin{pmatrix} \lambda_1 & 1 & 0 & 0 \\ 0 & \lambda_1 & -1 & 0 \\ 0 & 0 & \lambda_1 & 0 \\ 0 & 0 & 0 & \lambda_2 \end{pmatrix}$$

or

$$T^{ij} = -e^i_1 k^j - k^i e^j_1 + \lambda_1 (k^i l^j + l^i k^j - e^i_1 e^j_1) - \lambda_2 e^i_2 e^i_2 .$$

The above classification is complete in the following sense:

Theorem 3

The symmetric tensors T^i_j and T'^i_j can be transformed onto each other by means of the Lorentz transformation Λ^i_j

$$T'^{ij} = \Lambda^i_k \Lambda^j_l T^{kl},$$

if and only if they belong to the same class of the classes I, C, II_+, II_-, III and their corresponding eigenvalues are equal to each other.

The Plebański classification [42] is more detailed. In the Plebański notation, we give the multiplicities of particular eigenvalues in the decomposition of the characteristic polynomial into the prime factors (neglecting 1), and also the letters T, C, N, S, in particular if these subspaces are timelike, complex, null and spacelike, respectively. We give the dimensions of the maximal Jordan cages corresponding to these eigenvalues in round brackets as indices. Thus, the particular classes decompose into the Segré and Plebański types in the following way:

Class	Segré type	Plebański type
I	[1, 1, 1, 1]	$[T-S_1-S_2-S_3]_{(1111)}$
	[(1, 1) 1, 1]	$[2T-S_1-S_2]_{(111)}$
	[(1, 1) 1, 1]	$[T-2S_1-S_2]_{(111)}$
	[(1, 1)(1, 1)]	$[2T-2S]_{(11)}$
	[(1, 1, 1) 1]	$[3T-S]_{(11)}$
	[(1, 1, 1) 1]	$[T-3S]_{(11)}$
	[(1, 1, 1, 1)]	$[4T]_{(1)}$
C	$[z, \bar{z}, 1, 1]$	$[C-C-S_1-S_2]_{(1111)}$
	$[z, \bar{z}(1, 1)]$	$[C-C-2S]_{(111)}$
II	[2, 1, 1]	$[2N-S_1-S_2]_{(211)}$
(II_+ and II_-)	[2 (1, 1)]	$[2N-2S]_{(21)}$
	[(2, 1) 1]	$[3N-S]_{(21)}$
	[(2, 1, 1)]	$[4N]_{(2)}$
III	[3, 1]	$[3N-S]_{(31)}$
	[(3, 1)]	$[4N]_{(3)}$

The whole classification described above concerned an arbitrary symmetric tensor. However, apart from the fact that it is symmetric, enegy-momentum tensor must satisfy the so-called energy conditions. Namely, let u^i be a normalized timelike vector directed into the future, $u^i u_i = 1$. We can imagine that this vector represents the four-velocity of a certain observer. Then $P^i = T^i_j u^j$ represents the density of the four-momentum in this observer's system. If the vector $P^i \neq 0$, it should be a vector directed into the future: timelike or null. This condition is satisfied if the following two inequalities are met:

$$T_{ij} u^i u^j \geqslant 0$$

and

$$T_{ij} u^j T^i_k u^k \geqslant 0$$

for all timelike vectors (and, in view of the continuity argument, null vectors) u^i. The first inequality represents the non-negativity of energy, the second one ensures that P^i is not a spacelike vector.

It appears that a symmetric tensor belonging to classes C, II_- and III cannot satisfy the energy conditions. Therefore, the energy-momentum tensor should belong to class I or II_+. Moreover, if it belongs to class I, its eigenvalues must satisfy the inequalities

$$\lambda_0 \geqslant |\lambda_1|, \quad \lambda_0 \geqslant |\lambda_2|, \quad \lambda_0 \geqslant |\lambda_3|,$$

while, if it belongs to class II_+, similarly,

$$\lambda_0 \geqslant |\lambda_1|, \quad \lambda_0 \geqslant |\lambda_2|.$$

It would be interesting to see to which types the most frequently used energy-momentum tensors correspond. In the case of the perfect fluid,

$$T^{ij} = (c^2\varrho + p)u^i u^j - p g^{ij},$$

we have class I, the eigenvalues $\lambda_0 = c^2\varrho$, $\lambda_1 = \lambda_2 = \lambda_3 = -p$ and the Plebański type $[T-3S]_{(11)}$. In a special case, $p = -c^2\varrho$, the energy-momentum tensor has the form of the cosmological term and belongs to type $[4T]_{(1)}$.

For the electromagnetic field

$$4\pi T^{ij} = -F^{ik}F^j_k + \frac{1}{4} g^{ij}F_{kl}F^{kl},$$

we should distinguish between the general case and that of a plane wave. In the general case, there is a reference system, namely the orthonormal basis (e_0, e_1, e_2, e_3), where the electric and magnetic fields are parallel $\mathbf{E}||B||\mathbf{e}_1$. We can readily calculate that in this system

$$T^{ij} = \frac{1}{8\pi}(E^2 + B^2)(e_0^i e_0^j - e_1^i e_1^j + e_2^i e_2^j + e_3^i e_3^j).$$

This energy-momentum tensor belongs to class I, type $[2T-2S]_{(11)}$, while the eigenvalues are $\lambda_0 = \lambda_1 = -\lambda_2 = -\lambda_3 = \frac{1}{8\pi}(E^2 + B^2)$; they are thus equal to $\pm$ the energy density.

The tensor of the electromagnetic field of the plane wave type has the form

$$F_{ij} = k_i e_j - e_i k_j,$$

where k^i is a null vector and e^i is a spacelike vector orthogonal to k^i. We can normalize the vector e^i in any way; choosing $e_i e^i = -4\pi$, we obtain

$$T^{ij} = k^i k_j,$$

thus, it is a tensor of class II_+, type $[4N]_{(2)}$, whose eigenvalues are all equal to 0.

Before we pass on to the classification of the Weyl tensor, let us discuss the Hodge dualization of external forms. In an oriented Minkowski space, we can assign to each p-form ω, which has in the "right handed" basis the components $\omega_{i_1 \ldots i_p}$, a $(4-p)$-form $*\omega$ with components given by the formula

$$*\omega_{i_{p+1} \ldots i_4} = \frac{1}{p!} \varepsilon_{i_1 \ldots i_4} \omega^{i_1 \ldots i_p}.$$

This operation is determined correctly, i.e. it does not depend on the choice of the "right handed" basis. Equivalently, we can say that the Hodge dualization commutes with the action of linear transformations with a positive determinant on the forms. This operation has a number of interesting properties. Firstly, it is a linear isomorphism of the space of p-forms onto that of $(4-p)$-forms. Secondly, it changes the sign of the scalar product of the forms:

$$\frac{1}{(4-p)!} *\omega_{i_{p+1}\dots i_4} *\tau^{i_{p+1}\dots i_4} = -\frac{1}{p!}\,\omega_{i_1\dots i_p}\,\tau^{i_1\dots i_p}.$$

Thirdly, for forms of degree 0, 2 and we have 4, $**\omega = -\omega$; and for forms of odd degree $**\omega = \omega$.

We are interested in real and complex two-forms. We call the complex two-form ω self-dual if $*\omega = -\mathrm{i}\omega$, and antiself-dual if $*\omega = \mathrm{i}\omega$. We can uniquely represent any complex two-form ω as the sum of a self-dual form ω^+ and an antiself-dual one ω^-, where

$$\omega^{\pm} = \frac{1}{2}\,(\omega \pm \mathrm{i}*\omega).$$

If the two-form ω is real, then $\omega^- = \overline{\omega}^+$; therefore, there is a one-to-one correspondence between real and self-dual (or antiself-dual) two-forms. The spaces of self-dual and antiself-dual forms are orthogonal.

Acting on self-dual two-forms

$$\omega^+_{ij} \mapsto \Lambda^k_i \Lambda^l_j \omega^+_{kl},$$

the proper Lorentz transformation leaves them self-dual and preserves their scalar products. From the orthonormal basis (e^0, e^1, e^2, e^3) in the space dual to the Minkowski vector space, we can form an orthonormal basis in the space of self-dual two-forms (τ^1, τ^2, τ^3), given by the formula

$$\tau^a = \frac{1}{2}\,\varepsilon^{abc} e^b \wedge e^c + \mathrm{i} e^0 \wedge e^a,$$

where $a, b, c = 1, 2, 3$. Therefore, the space of self-dual two-forms is a three-dimensional Euclidean complex space. The Lorentz transformation $e'^i = \Lambda^i_j e^j$ induces in this space the orthogonal transformation $\tau'^a = O^a_b \tau^b$, where

$$O^a_b = \Lambda^0_0 \Lambda^a_b - \Lambda^0_b \Lambda^a_0 - \mathrm{i}\varepsilon^{acd} \Lambda^c_0 \Lambda^d_b.$$

The map $\Lambda \mapsto O$ is a homomorphism of the proper Lorentz group $\mathbf{L}_+$ onto the group $\mathbf{SO}(3, \mathbf{C})$ of complex orthogonal transformations with unit determinant. Moreover, the group $\mathbf{L}_+^{\uparrow}$ is isomorphic to $\mathbf{SO}(3, \mathbf{C})$.

Since the Weyl tensor is skew-symmetric in two pairs of indices, we can consider its left and right Hodge dualizations:

$$*C^{ijkl} = \frac{1}{2}\,\varepsilon^{ijmn} C_{mn}{}^{kl},$$

$$C*^{ijkl} = \frac{1}{2}\,\varepsilon^{klmn} C^{ij}{}_{mn}$$

and the left and right self-dual tensors. We can also perform similar operations on the Riemann tensor, however, in contrast to the Riemann vector, the Weyl vector has a certain special property, namely

$$*C^{ijkl} = C*^{ijkl},$$

from which it follows that the left and right self-dual tensors are equal to each other:

$$^{+}C^{ijkl} = C^{+ijkl}.$$

Since this is so, the mapping $Q:\omega^{+}_{ij} \mapsto \frac{1}{2}C^{+kl}_{ij}\omega^{+}_{kl}$ maps the space of self-dual two-forms into itself. Since it fully describes the Weyl tensor, the eigenproblem in the form

$$\frac{1}{2}C^{+kl}_{ij}\omega^{+}_{kl} = \lambda\omega^{+}_{ij}$$

leads to its classification. We can write it in the three-dimensional form

$$Q^{a}_{b}\omega_{a} = \lambda\omega_{b},$$

where the matrix Q^{a}_{b}, representing the mapping Q in the basis (τ^{a}), has the form

$$Q^{a}_{b} = -\frac{i}{2}\varepsilon^{acd}C^{+cd}{}_{0b} + C^{+0a}{}_{0b}.$$

Thus, we have assigned to the Weyl tensor a symmetric and trace-free mapping of the space of complex self-dual two-forms into itself. The correspondence is one-to-one. The classification of the mapping Q gives the sought classification of the Weyl tensor.

The mapping Q may lead to the following (complex) Segré types: [1, 1, 1], [(1, 1) 1], [2, 1], [(2, 1)], [3]. Let us note that it follows from the trace-free condition that only $Q^{a}_{b} = 0$, i.e. $C^{i}{}_{jkl} = 0$ corresponds to type [(1, 1, 1)]; in this case, we say that we are dealing with type 0.

The classification of the Weyl tensor is particularly interesting, if we represent it in the language of two-component spinors. Let us first consider the spinor image of the Riemann tensor

$$R^{ABCD\dot{E}\dot{F}\dot{G}\dot{H}} = \sigma_{i}{}^{A\dot{E}}\sigma_{j}{}^{B\dot{F}}\sigma_{k}{}^{C\dot{G}}\sigma_{l}{}^{D\dot{H}}R^{ijkl}.$$

The symmetries of the Riemann tensor lead to the reduction of independent components of this spinor.

First of all, taking into account the antisymmetry of R^{ijkl} in the first and the last pairs of indices and the fact that it is real, we obtain, similarly as for the electromagnetic field, the result:

$$R^{ABCD\dot{E}\dot{F}\dot{G}\dot{H}} = X^{ABCD}\varepsilon^{\dot{E}\dot{F}}\varepsilon^{\dot{G}\dot{H}} + \overline{X}^{\dot{E}\dot{F}\dot{G}\dot{H}}\varepsilon^{AB}\varepsilon^{CD} + \Phi^{AB\dot{G}\dot{H}}\varepsilon^{CD}\varepsilon^{\dot{E}\dot{F}} + \overline{\Phi}^{\dot{E}\dot{F}CD}\varepsilon^{AB}\varepsilon^{\dot{G}\dot{H}},$$

where the spinors X^{ABCD} and $\Phi^{AB\dot{G}\dot{H}}$ are symmetric in the first and the last pairs of indices. The symmetry with respect to the transposition of the pairs, $R^{ijkl} = R^{klij}$, gives two conclusions. Firstly, the spinor $\Phi^{AB\dot{G}\dot{H}}$ is Hermitian:

$$\Phi^{AB\dot{G}\dot{H}} = \overline{\Phi}^{\dot{G}\dot{H}AB}.$$

Secondly,

$$X^{ABCD} = X^{CDAB},$$

this, in turn, implies the following form of the spinor X^{ABCD} :

$$X^{ABCD} = \Psi^{ABCD} + \Omega^{BC}(\varepsilon^{AD}\varepsilon^{CA} + \varepsilon^{DB}\varepsilon^{DB}),$$

where we denoted

$$\overline{\Psi}^{ABCD} = X^{(ABCD)}.$$

The quantity $R^{i[jkl]}$ written in the spinor image is a spinor built from ε's multiplied by $\Omega - \overline{\Omega}$. Its vanishing thus gives one condition: $\overline{\Omega} = \Omega$.

Finally, we have three quantities describing the Riemann tensor: Ω, $\Phi^{AB\dot{G}\dot{H}}$ and Ψ^{ABCD}. A representation of the group **SL** (2, **C**) acts irreducibly on each of them. The scalar Ω is expressed by the curvature scalar $\Omega = \frac{1}{24}R$. The Hermitian and symmetric spinor $\Phi^{AB\dot{G}\dot{H}}$ is the spinor image of the trace-free Ricci tensor

$$R^{ij} - \frac{1}{4}g^{ij}R,$$

while, the full symmetric spinor Ψ^{ABCD} describes the Weyl tensor, namely, the spinor image of the Weyl tensor is given by the formula

$$C^{ABCD\dot{E}\dot{F}\dot{G}\dot{H}} = \Psi^{ABCD}\varepsilon^{\dot{E}\dot{F}}\varepsilon^{\dot{G}\dot{H}} + \overline{\Psi}^{\dot{E}\dot{F}\dot{G}\dot{H}}\varepsilon^{AB}\varepsilon^{CD}.$$

An interesting fact is that in the case of an empty spacetime, the Bianchi identity reduces to

$$\nabla_{A\dot{E}}\Psi^{ABCD} = 0,$$

and thus to a covariant generalization of the equation of the fields of massless particles with spin 2.

The principal directions of the spinor Ψ^{ABCD}, which give rise to the principal directions of the Weyl tensor, are determind in the following way:

$$\Psi^{ABCD} = \alpha^{(A}\beta^{B}\gamma^{C}\delta^{D)}.$$

The Weyl tensor is of the general type, also called *Petrov type* I, if none of its principal directions coincide. In the opposite case, we say that the Weyl tensor is algebraically special. Depending on the way in which the principal directions coincide, we distinguish the following four Petrov types: II, D, III and N. These coincidences and the corresponding Segré type of the mapping Q

are given in the table. The classification of the Weyl tensor given here was known to E. Cartan [5] already in 1922, Independently, it was discovered by A. Z. Petrov [41]; in the form represented here we owe it to R. Penrose [39].

Petrov type	Coincidences of principal directions	Segré type of mapping Q
I		[1, 1, 1]
II		[2, 1]
D		[(1, 1) 1]
III		[(2, 1)]
N		[3]
0	—	[(1, 1, 1)]

The algebraic classification of the Weyl tensor is of particular significance for the problem of gravitational radiation. Here, we have an analogy with the electromagnetic field. That is, the field of electromagnetic radiation far away from the source takes the shape of a plane wave, More specifically, if r is the distance from the sources, then for $r \to \infty$ (when the advanced time is kept fixed) the tensor (or spinor) F of the electromagnetic field has the asymptotic form

$$F = \frac{\mathrm{N}}{r} + \frac{\mathrm{I}}{r^2} + O\left(\frac{1}{r^3}\right),$$

where N and I denote the algebraically special and general types, respectively. Far away from the sources the principal directions of the electromagnetic field coincide, while close to them they separate.

A similar phenomenon occurs for the gravitational field. Here, the asymptotic behaviour of the Weyl tensor is determined by the so-called "peeling off" theorem:

$$C = \frac{\mathrm{N}}{r} + \frac{\mathrm{III}}{r^2} + \frac{\mathrm{II}}{r^3} + \frac{\mathrm{G}}{r^4} + \frac{\mathrm{I}}{r^5} + O\left(\frac{1}{r^6}\right).$$

In this formula, the symbols of the types are the same as those in the table; moreover, G denotes a field of type I where all the principal directions are geodesic.

CHAPTER 15

A Review of Phenomena Predicted by the Einstein Theory of Gravitation

The views which, in the present-day understanding, make up the general relativity theory, may be arranged in three groups.

1. The assumption that spacetime is a four-dimensional differential Riemannian manifold with the metric tensor signature $(+, -, -, -)$. The interpretation of the length of the arc s as time measured by ideal clocks. The interpretation of the timelike geodesics as the world-lines of free falls, and that of null geodesics as light rays. Moreover, we have the "principle of minimal gravitational coupling", which establishes the form of the equations of fields or equations of particles in the general theory of relativity.

According to this principle, we need to take equations which are valid in the special theory of relativity, write them covariantly (i.e. so that their form does not depend on the choice of the reference system), and then assume the same form of the equations in the general theory of relativity. The construction of dynamical equations according to this recipe is sometimes ambiguous. Let us take, e.g. the Maxwell equations

$$\nabla_j F^{ij} = -\frac{4\pi}{c} j^i, \quad \nabla_j F_{jk} + \nabla_k F_{ij} + \nabla_j F_{ki} = 0.$$

In the special theory of relativity these equations are equivalent to the following equations for the four-potential:

$$\Box A^i = \frac{4\pi}{c} j^i, \quad \nabla_i A^i = 0,$$

but in curved spacetime these two sets of equations are not equivalent. In the general theory of relativity it is assumed that the electromagnetic field is described by the first system of equations because, among other things, this system results from an appropriate variational pinciple.

2. The *principle of general invariance*, which can be expressed in the following form: the field g_{ji} is sufficient to describe the spacetime relations and the

gravitational field, and is subject to dynamical equations. Accordingly, apart from spacetime itself, there are no absolute elements.

3. The Einstein equations

$$G_{ij} = \varkappa T_{ij}.$$

We can draw some conclusions just on the basis of the first group of assumptions. They include the form of the metric for a weak gravitational field, obtained in the previous chapter,

$$g_{00} = 1 + \frac{2\varphi}{c^2}.$$

On the same basis, we obtain the "gravitational redshift", i.e. the phenomenon of change in the frequency of light propagating in the gravitational field.

In keeping with the postulates of the first group, light propagates along null geodesics. How do we find them? Let Ψ satisfy the eikonal equation

$$g^{ij} \frac{\partial \Psi}{\partial x^i} \frac{\partial \Psi}{\partial x^j} = 0,$$

and let

$$\frac{\partial \Psi}{\partial x^i} \neq 0.$$

Then, the curve $\lambda \mapsto x^i(\lambda)$ satisfying the equation

$$\frac{\mathrm{d}x^i}{\mathrm{d}\lambda} = g^{ij} \frac{\partial \Psi}{\partial x^j}$$

is a null geodesic contained within the hyperplane $\Psi = \text{const}$.

Let us take a stationary field g_{ij}, i.e. such a field that in a certain reference system

$$\frac{\partial g_{ij}}{\partial t} = 0,$$

where $ct = x^0$. In the stationary field there are solutions of the eikonal equation in the form

$$\Psi(x^i) = -\Omega t + \Phi(x^\alpha).$$

Let us imagine two observers "at rest", i.e. those for whom $x^1, x^2, x^3 = \text{const}$. Let observer 1 emit electromagnetic waves to observer 2 (Fig. 15.1). The surfaces of constant Ψ are the surfaces of the constant phases of these electromagnetic waves. The time interval

$$\Delta s = \int \mathrm{d}s = c \int \sqrt{g_{00}}\, \mathrm{d}t = c\sqrt{g_{00}} \int \mathrm{d}t = c\sqrt{g_{00}}\, \Delta t$$

between the same two phases measured by observers 1 and 2 is different, if g_{00} has different values at the points of spacetime where these observers are.

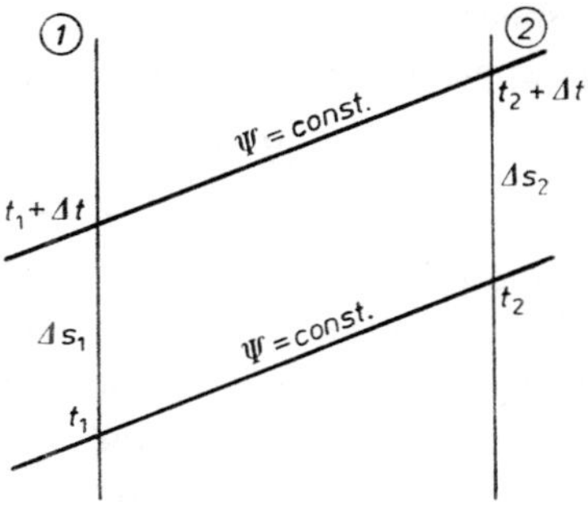

Fig. 15.1

Accordingly, the frequencies of the electromagnetic wave measured by these observers are different. Their ratio is

$$\frac{\omega_2}{\omega_1} = \frac{\Delta s_1}{\Delta s_2} = \frac{\sqrt{g_{00}|_1}}{\sqrt{g_{00}|_2}}.$$

For weak fields, since $g_{00} = 1 + (2\varphi/c^2)$, we obtain

$$\frac{\omega_2}{\omega_1} = 1 + \frac{\varphi_1 - \varphi_2}{c^2}.$$

For example, if 1 is a star or the Sun and 2 is the Earth, we can neglect the potential φ_2, since it is small as compared with φ_1, and, since $\varphi_1 < 0$—as it is the potential of an attractive force—we obtain

$$\omega_2 < \omega_1.$$

This is the phenomenon called the "redshift". The phenomenon of change in the frequency of electromagnetic waves in the Earth's gravitational field has been confirmed by means of the Mössbauer effect. It was found to agree with the above formula with an accuracy up to 10% [44].

In the Newtonian theory of gravitation, for a spherically symmetric body with mass M, we have the gravitational potential

$$\varphi = -\frac{kM}{r},$$

therefore, in the general theory of relativity for the same situation we should expect

$$g_{00} \cong 1 - \frac{2kM}{rc^2}.$$

It turns out that the unique solution of the Einstein equation for the empty space (outside the body)

$$G_{ij} = 0,$$

on the assumption of spherical symmetry, is the *Schwarzschild spacetime*, whose metric form in an appropriate coordinate system is

$$ds^2 = c^2\left(1-\frac{2kM}{rc^2}\right)dt^2-\left(1-\frac{2kM}{rc^2}\right)^{-1}dr^2-r^2(d\theta^2+\sin^2\theta d\varphi^2).$$

Schwarzschild [48] found this solution as ealy as 1916. Just as in the Newtonian theory, this solution depends exclusively on the mass of a body and not on its (spherically symmetric) structure; moreover, it is always static—even if the body undergoes spherically symmetric oscillations.

In the Schwarzschild spacetime a special role is played by a null hypersurface given by the equation $r = r_g$, where $r_g = 2kM/c^2$ is the so-called gravitational radius of the body. It was believed for some time that the geometry of spacetime on this hypersurface was singular. This did not lead to essential difficulties, since it was believed for a long time that for all bodies the gravitational radius is much smaller than the geometric radius, and therefore the apparent singularity of the metric at $r = r_g$ does not play any role at all (Fig. 15.2). As an example, we shall give the gravitational radii for a few bodies:

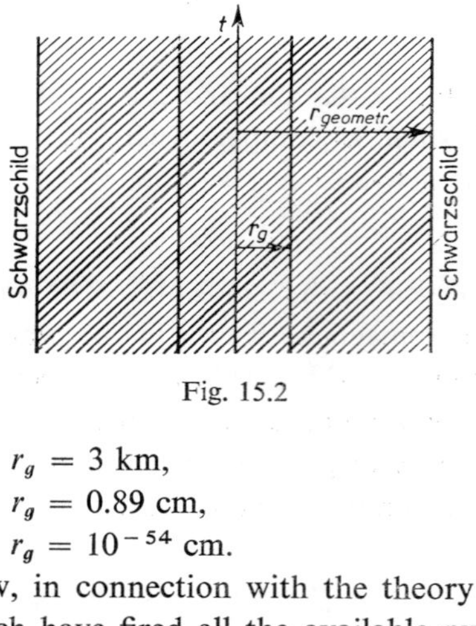

Fig. 15.2

for the Sun $r_g = 3$ km,
for the Earth $r_g = 0.89$ cm,
for a neutron $r_g = 10^{-54}$ cm.

For some time now, in connection with the theory of the late phases of evolution of stars which have fired all the available nuclear fuel, researchers have admitted the existence of material objects with dimensions of the same order of magnitude as those of their gravitational radii.

To verify that the surface $r = r_g$ is not—as it would seem—singular, but null, it suffices to replace the time t by the time t' defined by the formula

$$c dt' = c dt-\left(1-\frac{2kM}{rc^2}\right)^{-1}dr,$$

leaving, the other coordinates r, θ, φ unchanged.

The light cone is tangent to the Schwarzschild sphere, its past part lies on the external side and its future part is on the internal side (Fig. 15.3). Hence,

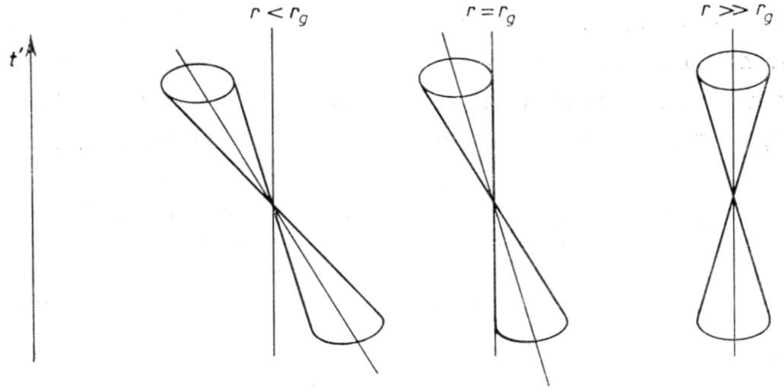

Fig. 15.3

it follows that particles can only enter the Schwarzschild sphere, but they cannot leave it. It turns out, moreover, that an observer on the outside can never receive signals notifying him that a particle has reached the sphere.

The Schwarzschild sphere can act as a trap for particles. In certain conditions, if the gravitational attraction dominates the pressure, the matter of a a star penetrates into the Schwarzschild sphere and the star "collapses"—this phenomenon is called *gravitational collapse* (Fig. 15.4).

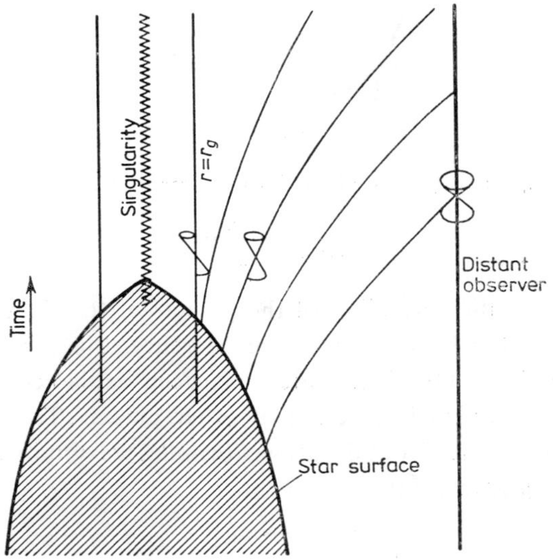

Fig. 15.4

According to present-day views, the gravitational collapse concerns all stars which have fully fired the nuclear fuel, and in the last stage of development have a mass exceeding a certain critical value, of the order of a few times the mass of the Sun.

This is connected with the problem of pulsars. It is thought that pulsars are rotating neutron stars which have fired all of their fuel, but their mass is too small for them to collapse. They are characterized by enormous densities and short radii, which means that the processes occurring in such stars should be significantly influenced by the effects of the general relativity theory.

	Sun	White dwarf (Sirius B)	Neutron star
Radius r	6.960×10^{10}cm	5.4×10^{8} cm	10^{6} cm
Density ϱ	1.410 g/cm^3	3.00×10^{6} g/cm^3	10^{14} g/cm^3
r_g/r	4×10^{-6}	5×10^{-4}	6×10^{-2}

On the basis of the Schwarzschild solution, we can predict other effects of the general theory of relativity, which include the motion of the perihelia of planets. In a spherically symmetric gravitational field, a particle moves (if its motion is bounded) along a rotating ellipse (Fig. 15.5). The advance of the

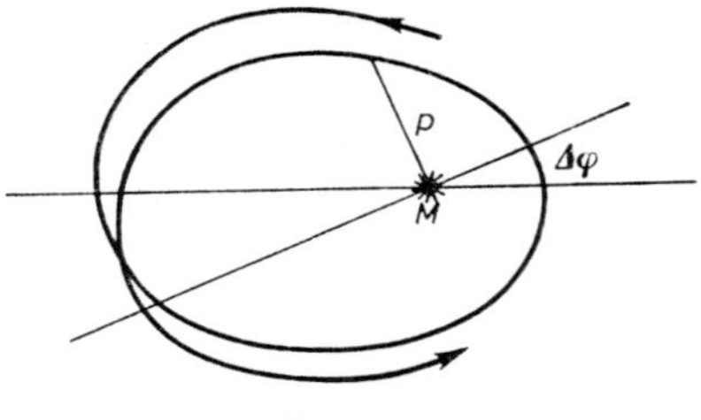

Fig. 15.5

perihelion after one rotation is given by the formula

$$\Delta\varphi = \frac{6\pi kM}{pc^2}.$$

This advance of the perihelion of the planets of the Solar System was observed as early as the 19th century. For Mercury, this advance is 1142″ per hundred years, about 1100″ of which has been explained by perturbations caused be the Newtonian interaction of other planets. To be accurate, 43.1″ ±0.4″ per century is left. From the above formula, we obtain 43.0″ per century, which is a striking agreement.

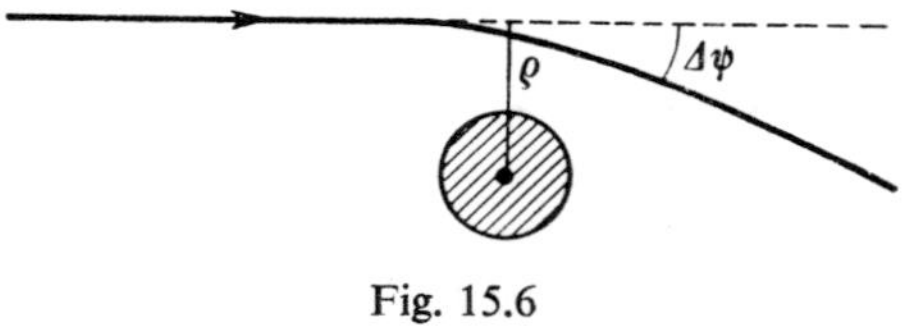

Fig. 15.6

Another effect, obtained by using the Schwarzschild solution, is the deflection of light rays close to stars (Fig. 15.6); we obtain here the result

$$\Delta\psi = \frac{4kM}{\varrho c^2}.$$

Substituting for M the mass of the Sun and replacing ϱ by its radius, we obtain $\Delta\psi = 1.75''$. The appropriate measurements, carried out during solar eclipses, involve quite large experimental error, so that they confirm the validity of the GRT only qualitatively, giving a result between $1''$ and $2.4''$ [2]. Since 1969 research has been carried out on the deflection by the Sun of radio waves coming from quasar $3C279$. The results of these measurements are much more accurate than the results of optical measurements, and they agree with the GRT predictions with an accuracy up to 10%.

In 1964, Shapiro [49] (Fig. 15.7) predicted another phenomenon, consist-

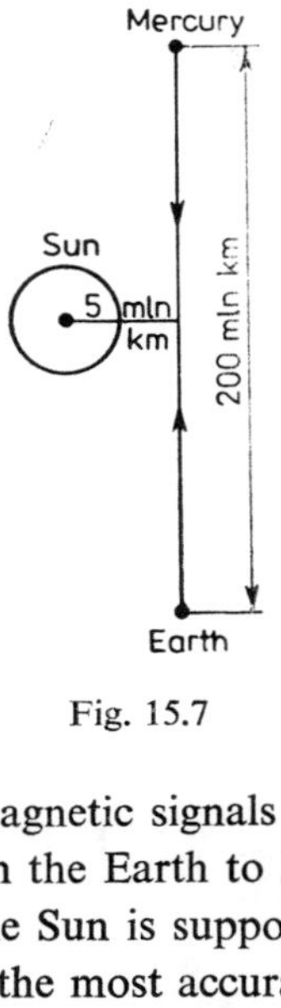

Fig. 15.7

ing in a time delay of electromagnetic signals passing close to the Sun. The delay of a signal travelling from the Earth to Mercury and back with respect to the passage time far from the Sun is supposed to be 200 μs. The satellites Mariner 6 and Mariner 7 gave the most accurate results. They agree with the Einstein-theory predictions with an accuracy up to 4%. Here, the Brans–Dicke theory gives a result which differs from the Einstein theory by 7%.

Observations of the deflection of light rays have shown that the Einstein theory is better than the Nordström theory, which assumed the validity of the first group of assumptions making up the general relativity theory, but rejected the arbitrariness of the metric tensor, assuming that the metric is conformally pseudo-Euclidean, i.e. in a certain coordinate system, it has the form

$$ds^2 = e^{\varphi}(c^2 dt^2 - dx^2 - dy^2 - dz^2),$$

and proposed, instead of the Einstein equations,

$$R = \varkappa T, \qquad \text{where} \qquad T = T_i^i.$$

This theory, however, failed to give the deflection of light rays, since the electrodynamic equations in vacuum are invariant with respect to conformal transformations.

CHAPTER 16

Gravitational Waves

Einstein validated theoretically the existence of gravitational radiation and gravitational waves soon after he had formulated the equations of the general theory of relativity [17]. Einstein's calculations at that time were based on approximate linear field equations and were often criticized for that reason; the linearization of the equations changes their properties significantly. Some conceptual difficulties in this field are related to the arbitrariness of coordinate systems. Taking the metric of a flat space in rectilinear coordinates and then carrying out a transformation of the form

$$x \mapsto x + f(x - ct),$$

we obtain a metric tensor with components looking as though they described wave motion. It appears therefore that, with an appropriate choice of the coordinate system, we can regard any gravitational field as a wave. Doubts arise from the fact that, because of its nontensorial nature, the gravitational force is not a well-defined notion, and neither is energy. However, very numerous and detailed studies on gravitational radiation, based on exact field equations and analyses of the invariant geometric properties of spacetime, have shown that Einstein's predictions were correct, confirming the equations which he obtained for the radiation power of isolated systems.

A large number of notions related to gravitational radiation are based on an analogy between electromagnetism and gravitation. The cornerstone of this analogy is the similarity between Coulomb's and Newton's laws, or, in other words, the fact that Poisson's equation is the basis of both electrostatics and the nonrelativistic theory of gravitation. The exclusively attractive nature of gravitational forces and the identity between the gravitational charge and the inertial mass are indications that we cannot take this analogy without qualifications.

Let us first consider a system of electric charges described in the special theory of relativity by the density $\varrho(\mathbf{r}, t)$. The solution of the wave equation for the scalar potential has the form

$$\varphi(\mathbf{r}, t) = \int \frac{1}{R} \varrho\left(\mathbf{r}', t - \frac{R}{c}\right) \mathrm{d}_3 x',$$

where $R = |\mathbf{r}-\mathbf{r}'|$. Let v, L and λ denote, respectively, the typical charge velocity, the linear dimensions of the region where charges move, and the typical length of electromagnetic waves generated by them. In the wave zone where $r \gg \lambda$, the distance R in the denominator of the integrated function can be replaced by r. In the nonrelativistic case $v \ll c$ and $L \ll \lambda$; therefore, we can replace R in the argument of ϱ by $r - \frac{\mathbf{r}\cdot\mathbf{r}'}{r}$, while we can expand ϱ into a power series in $1/c$. Then, after integration, we obtain

$$\varphi = \frac{e}{r} + \frac{\mathbf{r}\cdot\dot{\mathbf{d}}(t-r/c)}{cr^2} + \dots,$$

where

$$e = \int \varrho \,\mathrm{d}_3 x$$

is the total charge, and

$$\mathbf{d}(t) = \int \mathbf{r}\varrho(\mathbf{r}, t)\,\mathrm{d}_3 x$$

is the *dipole moment* of the system.

Although the scalar potential φ does not give full information about the electromagnetic field, since the essential part of information is contained in the vector potential, the above equation turns out to be sufficient for estimating the order of magnitude of the amount of radiated electromagnetic energy in a dipole approximation. In the wave zone, the electric field is of the order of $|\mathrm{grad}\,\varphi| \sim d(t-r/c)/c^2r + O(1/r^2)$, and the magnetic field is of the same order. This information is sufficient for estimating the energy radiated per unit time in the form of electric dipole waves, $P_e \sim \ddot{d}^2/c^3$. To be exact, we have

$$P_e = 2\ddot{\mathbf{d}}^2/3c^3.$$

Let us now consider the analogous wave equation

$$\Delta\varphi - \frac{1}{c^2}\frac{\partial^2\varphi}{\partial t^2} = 4\pi k\varrho$$

for the gravitational scalar potential φ generated by the system of masses described by the density ϱ. Making assumptions resembling those for the electromagnetic case, we reach the multipole expansion

$$\varphi = -k\left(\frac{m}{r} + \frac{\mathbf{r}\cdot\mathbf{p}}{cr^2} + \frac{x^\alpha x^\beta \ddot{Q}_{\alpha\beta}(t-r/c)}{2c^2r^3}\right) + \dots,$$

where

$$m = \int \varrho\,\mathrm{d}_3 x$$

is the total mass, while

$$Q_{\alpha\beta} = \int \left(x_\alpha x_\beta - \frac{1}{3}\,\delta_{\alpha\beta}\,r^2\right)\varrho\,\mathrm{d}_3 x$$

is the *tensor of the quadrupole moment*. To show that

$$\mathbf{p} = \int \dot{\varrho}\mathbf{r}\,d_3x$$

is the total momentum of the system, we need to use the continuity equation

$$\dot{\varrho} + \operatorname{div} \varrho \mathbf{v} = 0.$$

The three dots in the equation defining φ represent higher multipoles and also a term corresponding to the spherically symmetric moment of the order $2 \sim \int r^2 \varrho d_3 x$. Since both the mass and the momentum are preserved in the nonrelativistic limit, we should not expect the gravitational radiation to contain simple terms of the monopole type, or of the electric dipole type.

Because observations have shown that light rays are deflected when passing close to the Sun's surface, we know that the gravitational field should be tensorial in nature. Therefore, just as in electrodynamics, the scalar potential φ contains only part of the information about gravitational radiation. For example, it does not say anything about gravitational effects caused by the rotation of bodies. These effects are analogous to those related to a stationary magnetic field. Since the total angular momentum is conserved, we cannot expect to obtain, in the approximation of low velocities, gravitational radiation of the magnetic dipole type. The spherically symmetric moment $\int r^2 \varrho d_3 x$ would lead to a "scalar" wave corresponding to gravitons with null spin. Such waves occur in certain modifications of the Einstein theory, e.g. in the Brans–Dicke theory. The deflection of light rays gives evidence against such scalar additions, so we shall neglect them. Therefore, the quadrupole component gives the principal contribution to the energy radiated by a system of slowly moving bodies.

By means of simple dimensional analysis, we can show from the formula for φ that

$$P_g \sim \frac{c}{k} \oint (\nabla\varphi)^2 d\sigma \sim \frac{k}{c^5} \dddot{Q}^2_{\alpha\beta}.$$

Exactly, we obtain

$$P_g = \frac{k}{5c^5} \dddot{Q}^2_{\alpha\beta}.$$

We can now use the formulae for P_e and P_g to compare the amount of energy radiated in the form of electromagnetic and gravitational waves by a system of two bodies, moving along a curve close to a circle, which have charges e and $-e$ and equal masses m. If we limit our considerations to the orders of magnitude, denoting the radius of the motion and its angular velocity by a and ω respectively, we can write that

(1) in the electromagnetic case

$$d \sim ea, \quad \ddot{d} \sim ea\omega^2, \quad m\omega^2 a \sim e^2/a^2,$$

and thus,

$$P_e \sim mc^2\omega \left(\frac{e^2}{mc^2 a}\right)^{5/2};$$

(2) in the gravitational case

$$Q \sim ma^2, \quad \dddot{Q} \sim ma^2\omega^3, \quad m\omega^2 a \sim km^2/a^2,$$

and thus,

$$P_g \sim mc^2\omega \left(\frac{km}{c^2 a}\right)^{7/2}.$$

In both cases the amount of radiated power strongly depends on the dimensionless parameters

$$\frac{e^2}{mc^2 a} \quad \text{or} \quad \frac{km}{c^2 a}.$$

Their ratio, e^2/km^2, is of the order of 10^{42} for an electron. For an atom, $e^2/mc^2 a \sim e^4/\hbar^2 c^2 \sim 1/(137)^2$, while for an average binary star the ratio $km/c^2 a$ is extremely small. The exponent in the formula defining P_g is higher than that in the equation defining P_e, because of the quadrupole nature of gravitational radiation.

Large amounts of gravitational energy are radiated by binary stars with very close components and short rotation periods (e.g. the system *WZ Sagittae* with masses of 0.6 and 0.03 of that of the Sun and a period of 81 minutes). We can also expect considerable gravitational radiation during the nonsymmetric collapse of stars [60, 61]. It is also interesting to consider the problem of the gravitational radiation from a charge moving in an external magnetic field. We know that certain components of cosmic radiation are a result of the synchrotronic radiation of relativistic electrons in the magnetic field. Infeld and Róża Trautman [26] showed that in a nonrelativistic approximation the intensity of gravitational radiation for this problem is of the order of

$$P_g \sim \frac{Mc^2}{r/c}\left(\frac{v}{c}\right)^4 \frac{kM}{c^2 r}.$$

The order of the term v/c is lower for this radiation than in the case of bodies which interact only gravitationally. Because of the difficulties related to the notion of the external field in the general theory of relativity, however, the physical interpretation of this result must be exercised with great caution.

CHAPTER 17

Great Numbers. Gravitation versus Quantum Phenomena

Let us compare the electromagnetic and gravitational interactions in the atomic domain. Let us consider the hydrogen atom. The ratio of these interactions is

$$e^2 : kMm = 0.2 \times 10^{40},$$

where M is the proton mass, m is the electron mass and e is the charge of the electron.

Such great numbers are also obtained in considering the Universe as a whole: estimating its age and the number of particles which it contains. Let us recall the Hubble law

$$v = r/T,$$

relating the velocities of escaping galaxies, v, with their distance r (measured in light years). The constant T with the time dimension, whose value is

$$T = 10^{10} \text{ years},$$

characterizes the age of the Universe. In atomic time units, the value of this constant is of the order of

$$T : \frac{e^2}{mc^3} \sim 10^{40}.$$

If we denote by ϱ the mean mass density in the Universe, the number of particles in its observable part is of the order (assuming that a proton is a "unit" particle)

$$\varrho (cT)^3 : M \sim 10^{80}.$$

It is interesting to quote Dirac's opinion of the occurrence of such great numbers. Namely, he says that we shall not be able to explain such large constants in any reasonable future theory which would comprise atomic and gravitational phenomena. He therefore suggests that we should assume that with the passage of time these quantities change in value. It is thought that

$$e^2 : kMm \sim t,$$

which, assuming that $e = \text{const}$, M, $m = \text{const}$, which is reasonably well confirmed by experiments, gives

$$k \sim \frac{1}{t}.$$

Because of the difficulties in explaining the formation of stars and galaxies, the hypothesis of the decreasing gravitational constant appears attractive. The formalization of this hypothesis consists in replacing the constant k by a certain scalar field k which would satisfy certain equations and describe, along with the metric tensor g_{ij}, the gravitational phenomena. Jordan gave one special theory with a variable gravitational constant, and Brans and Dicke gave another. At present, attempts are being made to confirm these theories experimentally.

There is no doubt that gravitational phenomena, as any other, must have certain quantum foundations. The present-day classic Einstein theory of gravitation is certainly an approximation of a more exact theory taking into account the quantum nature of the microworld phenomena. Probably all physicists accept this generally formulated view. A large number of theoreticians also have positive views on how to construct a quantum theory of gravitation. The prevailling opinion is that it should be done following the pattern of electrodynamics, treating the metric as a potential and substituting certain components of this potential by operators satisfying appropriate commutation rules, etc. This procedure, although it encounters large technical difficulties, related to general invariance and the nonlinearity of equations, is essentially feasible. We should bear in mind, however, that not every classical theory should be "quantized". At any rate, we cannot do this with respect to statistical theories. It is difficult to believe, on the other hand, that the theory of gravitation should be a statistical, thermodynamic-type theory. The analogies between Einstein's theory of gravitation and the Maxwell electrodynamics are so striking that opponents of quantizing the general theory of relativity are a minority. We can expect, and this is confirmed by calculations performed by various authors, that the quantized theory of gravitation leads to effects resembling those obtained in the electrodynamics: transmutations of particles in the presence of gravitons, gravitational corrections to dispersion and to the energy levels of atoms, etc. We can readily foresee that these effect are enormously small. A quantity with the dimension of length

$$l = \sqrt{k\hbar/c^3} = 1.6 \times 10^{-33} \text{ cm}$$

plays the role of the coupling constant between the gravitational field and matter. The quantum gravitational effects are proportional to the corresponding powers of the ratio l/λ, where λ is the characteristic wavelength of the given problem. For energies now accessible, this ratio is very small.

It seems to us that if connecting the theory of relativity with quantum theory brings anything essentially novel into physics, this will happen in a different way than a conventional quantization of the gravitational field. We should bear in mind the fact that the theory of relativity is also a theory of the spacetime structure. It is underlain by the hypothesis that spacetime is continuous; more exactly, that it has the structure of a differential manifold. This assumption seems to be valid on the grounds of classical physics, but it is far from being a certainty, if we take into account the quantum nature of phenomena. Theories with continuous spacetime assume the possibility of identifying arbitrarily close events. But this is not possible, because of the atomic structure of matter and the finite dimensions of elementary particles. We believe that this essential impossibility should be reflected in the structure of spacetime just as the local indistinguishability between gravitational and inertial forces is in a natural way taken into account in the general theory of relativity. Of course, each change in the assumptions on the structure of spacetime would involve a thorough revision of the theory of gravitation and the whole of physics.

CHAPTER 18

Cosmology

Cosmology is a field of physics whose task is to describe the Universe as a whole. Cosmology neglects all local inhomogeneities and studies the geometry and motion of matter averaged over very large areas. The matter in the Universe is regarded as a fluid whose particles are clusters of galaxies. We assume that all sufficiently large regions of the Universe are the same in terms of the distribution and motion of matter. This statement, which has been confirmed by astronomical observations over the scale of 10^8 light years, is called the principle of spatial homogeneity of the Universe, and provides the basis of most cosmological models.

The mathematical counterpart to this principle is the action in spacetime of a certain Lie group G, whose orbits are spacelike hypersurfaces which fill the spacetime. All the geometrical quantities describing the state of matter, namely the metric tensor, connection, the four-velocity of matter, the energy density, pressure etc., should be invariant with respect to G.

Apart from the principle of spatial homogeneity of the Universe, we usually assume the principle of isotropy of space, which asserts that no direction is singled out in the Universe. This principle is well confirmed by observations, in particular by the isotropy of microwave cosmic radiation. This radiation fills the whole Universe, and its spectrum is that of the black body radiation at a temperature of 2.7 K.

The mathematical expression of the principle of isotropy of space is the action in spacetime of a Lie group G, such that the subgroup of isotropy G_p of any event p (i.e. the subgroup preserving p) contains an effectively acting group of rotations in three dimensions $\mathbf{SO}(3)$. Just as in the case of the principle of homogeneity of space, all the geometric quantities describing the behaviour of matter should be invariant with respect to G. The invariance of the metric tensor with respect to G leads to the statement that it should have the *Robertson–Walker form*

$$ds^2 = c^2 dt^2 - a^2(t) \frac{dx^2 + dy^2 + dz^2}{\left(1 + \frac{\varepsilon}{4}(x^2 + y^2 + z^2)\right)^2},$$

where $\varepsilon = 0, \pm 1$. The coordinate t is called *cosmic time*. The function $a(t)$

is called the *scale factor*. If $\varepsilon = 0$, the hypersurfaces of constant cosmic time have the metric of a flat space. If $\varepsilon = 1$, these hypersurfaces can be realized as three-dimensional spheres immersed in four-dimensional Euclidean space. If $\varepsilon = -1$, the above spatial metric is that of the three-dimensional hyperboloids immersed in the Minkowski space. We thus speak of a flat, a spherical and a hyperbolic world, respectively.

The principle of isotropy of space contains the principle of its homogeneity. In isotropic cosmology, the energy-momentum tensor must have the form of the energy-momentum tensor of the perfect fluid

$$T^{ij} = (c^2\varrho + p)u^i u^j - pg^{ij},$$

where $(u^i) = (1, 0, 0, 0)$, while the mass density ϱ and the pressure p are functions of the cosmic time t. The world-lines of matter are geodesics perpendicular to spacelike hypersurfaces (to the orbits of the group G), while the proper time of matter coincides with the cosmic time t.

When applied to the Robertson–Walker metric tensor, the Einstein equations take the form of the Friedmann equations

$$8\pi k\varrho = 3\frac{\dot{a}^2 + \varepsilon c^2}{a^2},$$

$$\frac{8\pi k}{c^2}p = -2\frac{\ddot{a}}{a} - \frac{\dot{a}^2 + \varepsilon c^2}{a^2},$$

where the dot denotes differentiation with respect to t.

Multiplying the second of these equations by 3 and adding it to the first one, we obtain

$$\frac{8\pi k}{c^2}(c^2\varrho + 3p) = -6\frac{\ddot{a}}{a}.$$

Since only those solutions of the Friedmann equations for which $\varrho > 0$ and $p \geqslant 0$ are of physical significance, it follows from the above equation that $\ddot{a} < 0$, and thus the function $a(t)$ is convex. Therefore, this function must vanish at one moment of time, at least. At that moment, ϱ becomes infinite. Thus, for any physically reasonable equation of state $\varrho = \varrho(p)$, the Einstein theory predicts that isotropic cosmological models should be singular, at least in the past.

To solve the system of Friedmann equations, we have to assume a certain equation of state. At the present stage of development of the Universe, the pressure is low compared with the mass density. Therefore, the assumption $p = 0$, i.e. the treatment of cosmic matter as dust, is reasonable. Then, the Friedmann equations have a first integral of mass conservation

$$\varrho a^3 = \text{const} = \frac{3}{4\pi}M,$$

where M is the mass contained in a "sphere" of volume $\frac{4}{3}\pi a^3$. The Friedmann equations now become

$$\frac{1}{2}\dot{a}^2 - \frac{Mk}{a} = -\frac{1}{2}\varepsilon c^2.$$

This equation has the form of the Newtonian law of energy conservation and is also valid in Newtonian cosmology, although it involves a different interpretation of the right-hand side from that in the relativity theory. It implies the expansion of the Universe in keeping with the Hubble law. Depending on ε, the solutions of this equation are:

(1) $\varepsilon = +1$:

$$ct = \frac{Mk}{c^2}(\eta - \sin\eta),$$

$$a = \frac{Mk}{c^2}(1 - \cos\eta).$$

(2) $\varepsilon = 0$:

$$a = \left(\frac{9}{2}Mk\right)^{1/3} t^{2/3}.$$

(3) $\varepsilon = -1$:

$$ct = \frac{Mk}{c^2}(\sinh\eta - \eta),$$

$$a = \frac{Mk}{c^2}(\cosh\eta - 1).$$

The dependence $a(t)$ is displayed in Fig. 18.1. Each of the three functions

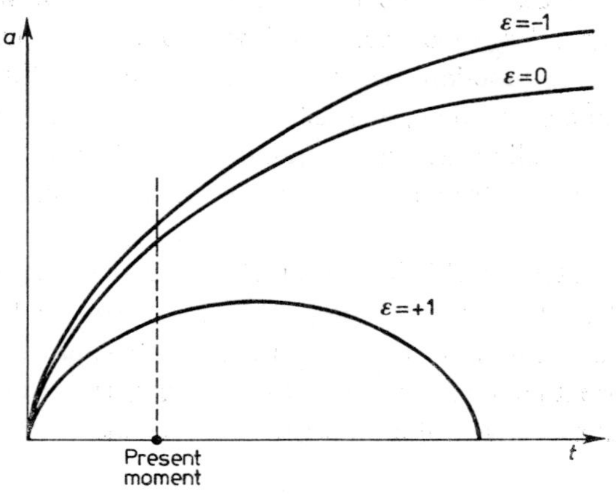

Fig. 18.1

vanished at least at one moment t, therefore we are dealing here with singular models. It has been believed for some time that it is possible to avoid singularities in more real cosmological models, taking into account local inhomogeneities and deviations from istropy. Exact studies, however, have shown that the existence of a singularity is a genaral property of solutions of the Einstein equations.

In the language of mathematics, the occurrence of singularities becomes manifest in the essential incompleteness of space time. A differential manifold with a linear connection is called geodesically complete, if each interval of any geodesic can be extended for arbitrarily large values of the affine parameter, i.e. if the geodesics have no beginning or end. It is easy to form an incomplete manifold by removing from a complete manifold, such as the Minkowski space, a certain closed set (e.g., a point). This incompleteness is unimportant, since we can remove it by returning the missing points. Singularities occurring in cosmology are essential, meaning roughly that models of the Universe have "holes" which cannot be filled.

In fact, in the Friedmann models discussed here, the lines $x, y, z =$ const. are geodesics determined for positive values of the affine parameter t. The metric and other physical parameters are also determined for $t > 0$; we cannot extend them smoothly to nonpositive moments of time, since the density of matter and the curvature scalar tend to infinity for $t \to +0$. Because of this, the initial period of very fast expansion of the Universe is called the *Big Bang*. The occurrence of a singularity means that within the classical Einstein theory of gravitation we cannot describe the very beginning of the explosion, or say what preceeded it.

Restrictions of the range of applicability occur in all known physical theories, and therefore we should not be surprised that the Einstein theory fails for very large densities and temperatures accompanying the early period of the Big Bang. The view is quite common that in describing correctly the initial—or rather "hot"—development stage of the Universe, we have to take into account quantum phenomena, in particular the quantum foundation of gravitation itself. It is supposed, however, that gravitational quantum phenomena—such as the creation of a pair of particles by gravitons—become manifest only for energies and curvatures corresponding to the Planck length (see Chapter 17), i.e. for moments of time comparable to

$$\sqrt{k\hbar/c} = 5.4 \times 10^{-44}\text{s}$$

and densities of matter of the order of

$$c^5/\hbar k^2 = 5.1 \times 10^{93}\text{g/cm}^3.$$

There are no reasons to suppose that the Einstein theory and other physical theories now known may be applied over such a wide range of energy and density. In other words, going back in time towards the Big Bang and the theoretical singularity, we are likely to come across new physical phenomena, which will undermine the significance of the Friedmann cosmological models, before we reach densities of the order of 10^{93} g/cm^3, characteristic of the creation of pairs by gravitons [62]. One of such phenomena—the effect of particle spin on geometry—was predicted by the Einstein–Cartan theory, described briefly in the next chapter. It turns out that within the Einstein–Cartan theory it is possible to build cosmological models without singularities [32]. According to these models, the density of matter at the moment of the greatest contraction of the Universe is of the order of (M—the proton mass)

$$M^2c^4/k\hbar^2 \simeq 10^{55}\text{g/cm}^3,$$

and, although it is high, it is much smaller than the characteristic density for gravitational quantum phenomena [55].

The superiority of the Einstein–Cartan theory over the other attempts of modification of general relativity consists, to our minds, in among other things, the fact that this theory was not constructed ad hoc, in order to eliminate the singularity. The nonsingular cosmological model with spin was found half a century after Cartan [6] had proposed his modification of the Einstein theory.

Incidentally, in considering the singularity, it is worthwhile to think about the evolution of a spherical closed model of the Universe. Taking literally—and formally—the equation of evolution of this model, we can represent the development of the Universe as a full cycloid (Fig. 18.2), corresponding to the

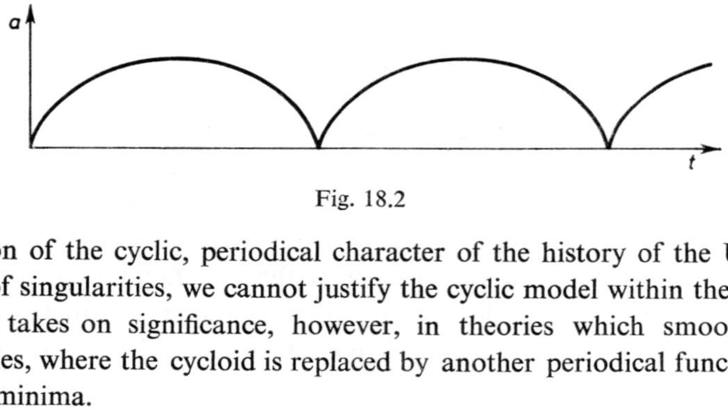

Fig. 18.2

recognition of the cyclic, periodical character of the history of the Universe. Because of singularities, we cannot justify the cyclic model within the Einstein theory; it takes on significance, however, in theories which smoothed out singularities, where the cycloid is replaced by another periodical function with spositive minima.

It is interesting to consider the question of whether the Universe is closed ($\varepsilon = 1$) or open ($\varepsilon = 0$ or $\varepsilon = -1$). We can answer the question by measuring the Hubble constant

$$H = \frac{\dot{a}}{a} = \frac{1}{T}$$

and the current matter density ϱ. The Universe is closed if ϱ exceeds the *critical density*

$$\varrho_c = \frac{3H^2}{8\pi k}.$$

Equivalently, the Universe is closed, if the deceleration parameter

$$q = -\frac{a\ddot{a}}{\dot{a}^2} = \frac{1}{2}\frac{\varrho}{\varrho_c}$$

is greater than $\frac{1}{2}$.

We can determine the three parameters ϱ, H (or ϱ_c)and q independently. The knowledge of their present-day values is unsatisfactory. The best known is the present-day value of H_0 of the Hubble "constant" $H(t)$. Its inverse, the *Hubble time*, is, according to observation data, $1/H_0 = (18 \pm 2) \times 10^9$ years. We should bear in mind, on the other hand, that this values has repeatedly been reduced, and so we cannot take the present value as final, either. This value of H_0 corresponds to the critical density $\varrho_{c0} \simeq 5 \times \times 10^{-30}$ g/cm³. The present-day value ϱ_0 of the density ϱ is little known, since it is difficult to take into account all the factors which affect it. The interval $\varrho_0 = 2 \div 6 \times 10^{-31}$ g/cm³ is taken as most likely. This would argue for the fact that the Universe is open. Calculations analyzing the dynamics of galaxies give, on the other hand, larger values of the mean density than simple estimation of their visible masses does. We call this the problem of missing mass. The greatest discrepancies occur for the present-day value q_0 of the deceleration parameter q. Estimations of q_0 indicate that $2q_0\varrho_{c0}$ is greater than the given value ϱ_0, which is probably related to the problem of missing mass. Due to cosmic neutrinos, the value ϱ_0 could be greater, if it turned out that neutrinos had mass. This would lead to the "closing" of the Universe.

In the above considerations, we neglected the cosmological term frequently introduced in the Einstein equations. The Einstein equations with the cosmological constant Λ have the form

$$R_{ij} - \frac{1}{2} R g_{ij} + \Lambda g_{ij} = \frac{8\pi k}{c^4} T_{ij}.$$

These equations have many properties in common with the proper Einstein equations. Above all, the generalized conservation law of energy momentum $\nabla_i T^{ij} = 0$ results from them. The basic difference is that if $\Lambda \neq 0$, the flat spacetime is not a solution of the vacuum equations. These equatins agree, with observations provided that Λ is sufficiently small. It is only then that the new term plays an essential role in cosmology; this remark gives it its name. The introduction of the cosmological term gives a much broader class of cosmological models. In theories with the cosmological constant, the observable parameters ϱ_0, H_0 and q_0 are independent, while the knowledge of them determines uniquely the model of the Universe.

CHAPTER 19

The Einstein-Cartan Theory

A few years after the general relativity theory had been proposed, Cartan [4, 6] suggested a certain modification of it, called today the Einstein–Cartan theory, consisting in rejecting the assumption of the symmetry of the linear connection. According to Cartan, the antisymmetric part of the connection should be related by algebraic equations with the spin tensor of physical fields generating gravitation.

We can give the following heuristic arguments for the Einstein–Cartan theory. In a description of elementary particles based on the special relativity theory, the principal role is played by the invariants of the Lie algebra of the Poincaré group, related to translations (mass) and rotations (spin). The invariance of the theory under translations ensures that the energy-momentum conservation law holds, just as the invariance of the theory under the Lorentz transformation ensures that the conservation law of angular momentum is satisfied. In the Einstein theory, mass is the cause of the curvature of spacetime, whereas spin has no direct influence on the geometry. Perhaps, this situation makes it difficult to understand the relation between the physics of elementary particles and the theory of gravitation. In the Einstein-Cartan theory, spin is the cause of the torsion of spacetime.

From the geometric point of view, the curvature is related to Lorentz transformations, whereas the torsion is connected with translations. To see this, let us consider the field of the radius vector r^i determined on the curve $x^i(t)$. It is defined by the formula

$$\frac{\mathrm{d}x^i}{\mathrm{d}t}\nabla_i r^j = \frac{\mathrm{d}x^j}{\mathrm{d}t}.$$

If the curve $x^i(t)$ is closed, the radius vector increases after completing a full cycle

$$\Delta r^i \cong (R^i{}_{jkl}\, r - Q^i{}_{kl})\Delta\tau^{kl},$$

where $\Delta\tau^{kl}$ is an oriented surface element spanned by this curve.

In the Einstein–Cartan theory, the connection is nonsymmetric, but it satisfies the metric condition

$$\nabla_i g_{jk} = 0.$$

The gravitational field equations, resulting from the appropriate variational principle, have the form

$$R_{ij}-\frac{1}{2}Rg_{ij} = \frac{8\pi k}{c^4}T_{ij},$$

$$Q^k{}_{ij}+\delta^k_i Q^l{}_{jl}-\delta^k_j Q^l{}_{il} = \frac{8\pi k}{c^3}S^k{}_{ij},$$

where T_{ij} denotes the asymmetric, so-called canonical, energy-momentum tensor. The second equation is a linear, reversible relation between spin and torsion. In the region of spacetime where spin vanishes, so does torsion, and the equations of the Einstein–Cartan theory turn into those of the Einstein theory [25, 54].

In the Einstein–Cartan theory, we can construct nonsingular cosmological models. In the simplest of these models [32], the metric is a plane Robertson–Walker metric

$$ds^2 = c^2dt^2-a^2(t)(dx^2+dy^2+dz^2).$$

The energy-momentum tensor has the dust form $T_{ij} = c^2\varrho u_i u_j$, $(u^i) = (1, 0, 0, 0)$. The spin tensor has the form $S^k_{ij} = u^k S_{ij}$, where $S_{12} = -S_{21}$ is the only nonvanishing component of the tensor S_{ij}. The nonvanishing component of the spin S_{12} singles out a certain direction in spacetime, therefore, although this model satisfies the principle of homogeneity of space, it is not an isotropic model. The spin obeys the conservation law

$$S_{12}\frac{4\pi}{3}a^3 = S = \text{const},$$

where S denotes the amount of spin contained in a sphere with a radius a. The scale factor a then satisfies the equation

$$\frac{1}{2}\dot{a}^2-\frac{Mk}{a}+\frac{3S^2k^2}{2c^4a^4} = 0.$$

In this equation, apart from the Newtonian term, we have an additional term playing the role of a "repulsive potential". This term occurs due to ordered spins. The solution of the above equation is

$$a = \sqrt[3]{\frac{9}{2}Mkt^2+\frac{3S^2k}{2Mc^4}}\,.$$

We can see that a does not vanish anywhere, therefore this cosmological model is nonsingular (Fig. 19.1). We obtain

$$a_{\min} = \sqrt[3]{\frac{3S^2k}{2Mc^4}}$$

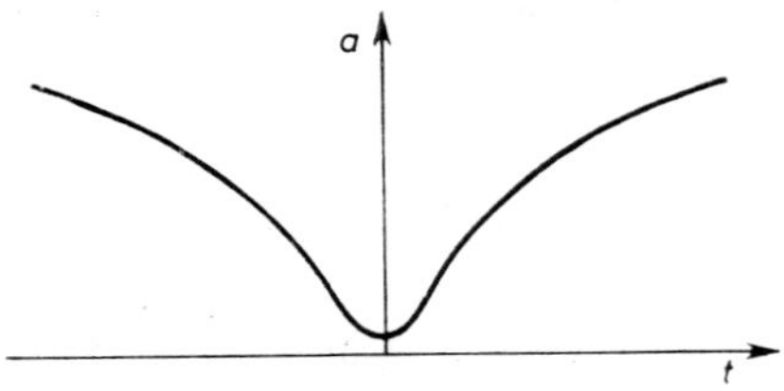

Fig. 19.1

as the minimum value of a. Substituting in this formula $M = m \times 10^{80}$, where m is the mass of a nucleon and 10^{80} is the number of nucleons in the part of the Universe accessible to observations and assuming the maximum ordering of spins, $S = \frac{1}{2}\hbar \times 10^{80}$, we obtain the upper limit of $a_{\min}$, $a_{\min} \cong 1$ cm [55]. We should point out that this short radius of the Universe is decreased further if we consider anisotropic metrics. Requiring that the metric should be nonsingular, we obtain a very strong restriction on the relative anisotropy of the Hubble constant,

$$\frac{\Delta H}{H} < 10^{-41}.$$

References

(A) References of general interest

Anderson J. L.: *Principles of relativity physics*, Academic Press, New York, 1967.

Bondi H.: *Cosmology*, 2nd Edition, Cambridge University Press, Cambridge, 1960.

Bondi H.: *Relativity and common sens*, Doubleday and Co., Garden City, 1964.

Demiański M.: *Relativistic astrophysics*, PWN, Warszawa, 1985.

Einstein A.: *The meaning of relativity*, 5th Edition, Princeton University Press, Princeton, 1956.

Fock V.: *The theory of space time and gravitation*, Pergamon Press, London, 1959.

Hawking S. W. and Ellis G.F.R.: *The large scale structure of space-time*, Cambridge University Press, Cambridge, 1973.

Hawking S. W. and Israel W. (editors): *General relativity: an Einstein centenary survey*, Cambridge University Press, Cambridge, 1979.

Held A. (editor): *General relativity and gravitation—one hundred yers after the birth of Albert Einstein*, vol. 1 and 2, Plenum Press, New York, 1980.

Kramer D., Stephani H., MacCallum M. and Herlt E. (Schmutzer E. editor): *Exact solutions of the Einstein field equations*, Deutscher Verlag der Wissenschaften, Berlin, 1980.

Landau L. D. and Lifshitz E. M.: *The classical theory of fields*, Pergamon Press, London, 1959.

Misner C. W., Thorne K. S. and Wheeler J. A.: *Gravitation*, W. H. Freeman and Co., San Francisco, 1973.

Pauli W.: *Relativitätstheorie*, in: *Encyklopädie der mathematischen Wissenschaften*, Vol. V19, B. G. Teubner, Leipzig, 1921.

Penrose R. and Rindler W.: *Spinors and Space-time*, vol. I and II, Cambridge University Press, Cambridge, 1984 and 1986.

Rindler W.: *Essential relativity*, revised 2nd Edition, Springer Verlag, New York, 1977.

Synge J. L.: *Talking about relativity*, North-Holland, Amsterdam, 1970.

Taylor E. F. and Wheeler J. A.: *Spacetime physics*, W. H. Freeman and Co., San Francisco and London, 1966.

Trautman A., Pirani P. A. E. and Bondi H.: *Lectures on general relativity*, Prentice-Hall, New Jersey, 1964.

Weinberg S.: *Gravitation and cosmology*, Wiley, New York, 1972.

Whittaker E.: *A history of the theories of eather and electricity*, Philosophical Library, New, York, 1954.

(B) References cited in the text

[1] Abraham M.: *Phys. Z.* **13** (1912) N. 1.

[2] Adam M. G.: The observational tests of gravitation theory, *Proc. Roy. Soc.* **A 270** (1962) 297.

[3] Bondi H., Pirani F. A. E. and Robinson I.: Gravitational waves in general relativity. III: Exact plane waves. *Proc. Roy. Soc.* **A 251** (1959) 519.

[4] Cartan E.: Sur une généralisation de la notion de courboure de Riemann et les éspaces à torsion, *C. R. Acad. Sci.* (Paris) **174** (1922) 593.

[5] Cartan E.: Sur les éspaces conformes généralisés et l'Univers optique, *C. R. Acad. Sci.* (Paris) **174** (1922) 857.

[6] Cartan E.: Sur les variétés à connexion affine et la théorie de la relativité généralisée, *Ann. Éc. Norm. Sup.* **40** (1923) 325; **41** (1924) 1.

[7] Cedarholm J. P. Bland G. P., Haveus B. L. and Townes C. H.: New experimental tests of specal relativity, *Phys. Rev. Lett.* **1** (1958) 342.

[8] Einstein A.: Zur Elektrodynamik der bewegter Körper, *Ann. Phys.* **17** (1905) 891.

[9] Einstein A.: Ist die Trägheit einer Körpers von seinem Einergiegehalt abhängig?, *Ann. Phys.* **18** (1905) 639.

[10] Einstein A.: Über das Relativitätsprinzip und die aus demselben gezogenen Folgerungen, *Jahrb. Radioakt. u Elektronik* **4** (1907) 411.

[11] Einstein A.: Über den Einfluss der Schwerkraft auf die Ausbreitung des Lichtes, *Ann. Phys.* **35** (1911) 898.

[12] Einstein A. (mit M. Grossmann): Entwurf einer verallgemeinerten Relativitätstheorie und Theorie der Gravitation, *Z. Math. u. Phys.* **62** (1913) 225.

[13] Einstein A.: Die formale Grundlage der allgemeinen Relativitätstheorie, *Sitz. Preuss. Akad. Wiss.* **47** (1914) 1030.

[14] Einstein A.: Zur allgemeinen Relativitätstheorie, *Sitz. Preuss. Akad. Wiss.* **44** (1915) 778.

[15] Einstein A.: Die Feldgleichungen der Gravitation, *Sitz. Preuss. Akad. Wiss.* **48** (1915) 844.

[16] Einstein A.: Die Grundlage der allgemeinen Relativitätstheorie, *Ann. Phys.* **49** (1916) 769.

[17] Einstein A.: Näherungsweise Integration der Feldgleichungen der Gravitation, *Preuss. Akad. Wiss. Sitz,* **1** (1916) 688.

[18] Einstein A.: Kosmologische Betrachtungen zur allgemeinen Relativitätstheorie, *Preuss. Akad. Wiss. Sitz.* **1** (1917) 142.

[19] Einstein A. and Grommer J.: Allgemeine Relativitätstheorie und Bewegungsgezetz, *Preuss. Akad. Wiss. Phys.-Math. Kl. Sitz.* (1927) 235.

[20] Einstein A. and Rosen N.: On gravitational waves, *Journ. Franklin Inst.* **223** (1937) 43.

[21] Einstein A., Infeld L. and Hoffmann B.: The gravitational equations and the problem of motion, *Ann. of Math.* **39** (1938) 65.

[22] Friedmann A. Über die Krümmung des Raumes, *Zs. f. Physik* **10** (1922) 377.

[23] Gamow G.: *Mister Tompkins in Wonderland*, Cambridge University Press Cambridge, 1953.

[24] Hawking S. W. and Penrose R.: The singularities of gravitational collapse and cosmology, *Proc. Roy. Soc.* **A 314** (1970) 529.

[25] Hehl F. W. von der Heyde P., Kerlick G. D. and Nester J. N.: General relativity with spin and torsion: Foundations and prospects, *Rev. Mod. Phys.* **48** (1976) 393.

[26] Infeld I. and Michalska-Trautman R.: Radiation from systems in nearly periodic motion, *Ann. Phys.* **55** (1969) 576.

[27] Infeld L. and Plebański J.: *Motion and Relativity*, PWN, Warszawa, 1960.

[28] Ives H. E. and Stillwell C. R.: Experimental study of the rate of a moving clock, *J. Opt. Soc. Amer.* **28** (1938) 215.

[29] Jabłoński A.: On models in Physics, *Postępy Fizyki* **20** (1969) 541.

[30] Kaluza T.: Zum Unitätsproblem der Physik, *Preuss. Akad. Wiss. Sitz.* (1921) 966.

[31] Kennedy R. J. and Thorndike E. M.: Experimental establishment of the relativity of time, *Phys. Rev.* **42** (1932) 400.

[32] Kopczyński W.: An anisotropic universe with torsion, *Phys. Lett.* **43A** (1973) 63.

[33] Larmor J. J.: *Aether and Matter*, Cambridge, 1900.

[34] Lorentz H. A.: De relative beweging van de aarde en dem aether, *Versl. gewone Vergad. Akad. Amst.* **1** (1982) 74.

[35] Michelson A. A.: *Light waves and their uses*, University of Chicago Press, Chicago, 1903.

[36] Minkowski H.: Raum and Zeit, *Phys. Z.* **10** (1909) 104.

[37] Nordström G.: Zur Theorie der Gravitation vom Standpunkt des Relativitätsprinzip, *Ann. Phys.* **42** (1913) 533.

[38] Penrose R.: The apparent shape of a relativistically moving sphere, *Proc. Cambridge Phil. Soc.* **55** (1959) 137.

[39] Penrose R.: A spinor approach to general relativity, *Ann. Phys.* (USA) **10** (1960) 171.

[40] Penzias A. A. and Wilson R. W.: *Astrophys. Journal* **142** (1965) 419.

[41] Petrov A. Z.: *Einstein spaces*, Pergamon Press, London, 1969.

[42] Plebański J.: The algebraic structure of the tensor of matter, *Acta Phys. Polon.* **26** (1964) 963.

[43] Poincaré H.: *Eclairage Electrique* **5** (1895) 5.

[44] Pound R. V., Rebka G. A. jr.: Apparent weight of photons, *Phys. Rev. Lett.* **4** (1960) 337.

[45] Recami E.: Theory of relativity and its generalizations, in: *Centenario di Einstein 1879–1970, astrofisica e cosmologia, gravitazione, quanti e relatività*, Giunti Barbéra, Firenze, 1979.

[46] Rindler W.: Length contraction paradox, *Amer. J. Phys.* **29** (1961) 365.

[47] Ritz W.: Recherches critiques sur l'électrodynamique générale, *Ann. Chim. Phys.* **13** (1908) 317.

[48] Schwarzschild K.: Über das Gravitationsfeld eines Massenpunktes nach der Einsteinschen Theorie, *Preuss. Akad. Wiss. Sitz.* (1916) 189.

[49] Shapiro I. I., Pettengill G. H., Ash M. E., Stone M. L., Smith W. B., Ingalls R. P. and Brockelman R. A.: Fourth test of general relativity: preliminary results, *Phys. Rev. Lett.* **20** (1968) 1265.

[50] Terrel J.: Invisibility of the Lorentz contraction, *Phys. Rev.* **116** (1959) 1041.

[51] Trautman A.: Galilean theory of relativity, *Postępy Fizyki* **17** (1966) 129.

[52] Trautman A.: Theory of relativity, *Postępy Fizyki* **17** (1966) 129.

[53] Trautman A.: Comparison of Newtonian and relativistic theories of space-time, in: *Perspectives in geometry and relativity*, Ed. B. Hoffman, Indiana Uinversity Press, 1966.

[54] Trautman A.: On the Einstein–Cartan equations, *Bull. Polon. Acad. Sci. sér. sci. math. astr. phys.* **20** (1972): Part I 185, Part II 503, Part III 895.

[55] Trautman A.: Spin and torsion may avert gravitational singularitities, *Nature Phys. Sci.* **242** (1973) 7.

[56] Voigt W.: Über das Dopplersche Prinzip, *Nachr. Ges. Wiss. Göttingen* (1887) 41.

[57] Weber J.: General relativity and gravitational waves, Wiley–Interscience, New York, 1961.

[58] Weisskopf V. F.: The visual appearance of rapidly moving objects, *Phys. Today* **13** (1960) 24.

[59] Weyl H.: Eine neue Erweiterung der Relativitätstheorie, *Ann. Phys.* **59** (1919) 101.
[60] Zel'dovich Ya. B. and Novikov I. D.: The radiation of gravitational waves moving in a field of a collapsing star, *DAN SSSR* **155** (1964), 1033.
[61] Shklovskii I. S. and Kardashov N. S.: The gravitational waves and superstars, *DAN SSSR* **155** (1964), 1039.
[62] Zel'dovich Ya. B. and Starobinskii A. A.: Creation of particles and vacuum polarization in an anisotropic gravitational field, *Zh. E. T. F.* **61** (1971) 2161.

Subject Index

Andrzej Trautman was born in Warsaw in 1933 and graduated from a high school in Paris in 1949. He prepared his Ph.D. thesis under the supervision of Leopold Infeld and obtained his doctorate in Warsaw in 1959. Since 1961 he has worked at Warsaw University and is now a full professor at the Institute of Theoretical Physics. Professor Trautman has been Research Fellow and Visiting Professor at King's College (1958) and Imperial College (1959–60), London, at Syracuse University (1961 and 1967), Collège de France (1963 and 1981), University of Chicago (1971), SUNY at Stony Brook (1976–77) and University of Montreal (1982 and 1990). Since 1983 he has been a frequent visitor to the International School for Advanced Studies and the International Centre for Theoretical Physics at Trieste. He is a member of the Polish and Czecho-Slovak Academies of Sciences and recipient of the Alfred Jurzykowski Prize in physics (New York, 1984). He has published several books: *Lectures on General Relativity* (with H. Bondi and F. A. E. Pirani, Prentice-Hall, 1965), *Differential Geometry for Physicists* (Bibliopolis, 1984), *The Spinorial Chessboard* (with P. Budinich, Springer, 1988). He has also written numerous articles on the theory of general relativity, applications of differential geometry to physics, classical Yang–Mills fields and spinors. He is married and has two sons.